LOOK INSIDE CROSS-SECTIONS SPACE

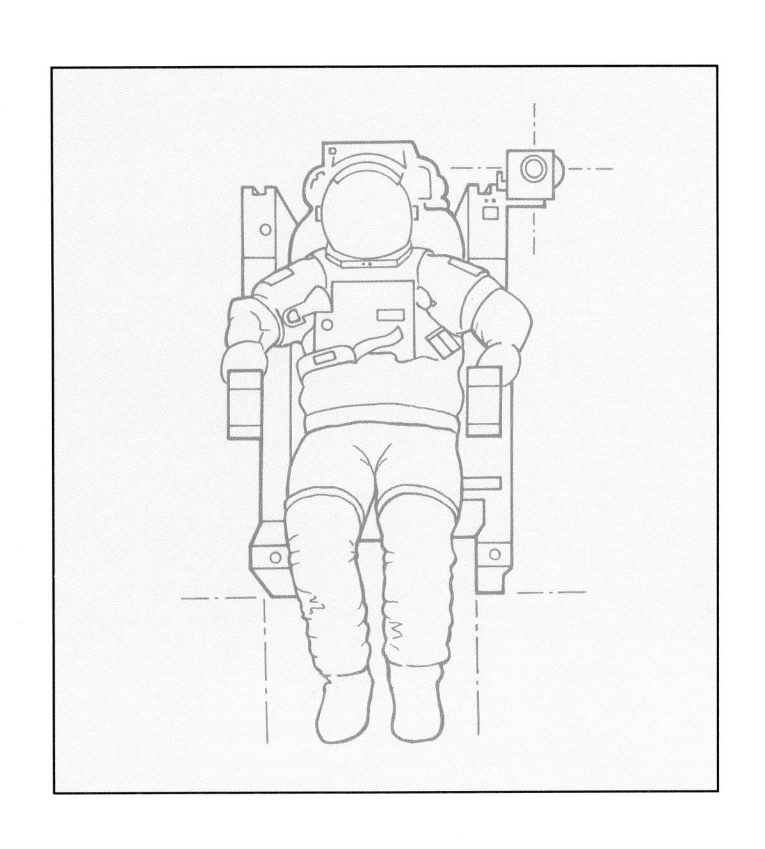

LOOK INSIDE

CROSS-SECTIONS

SPACE

ILLUSTRATED BY

NICK LIPSCOMBE AND GARY BIGGIN

WRITTEN BY

MOIRA BUTTERFIELD

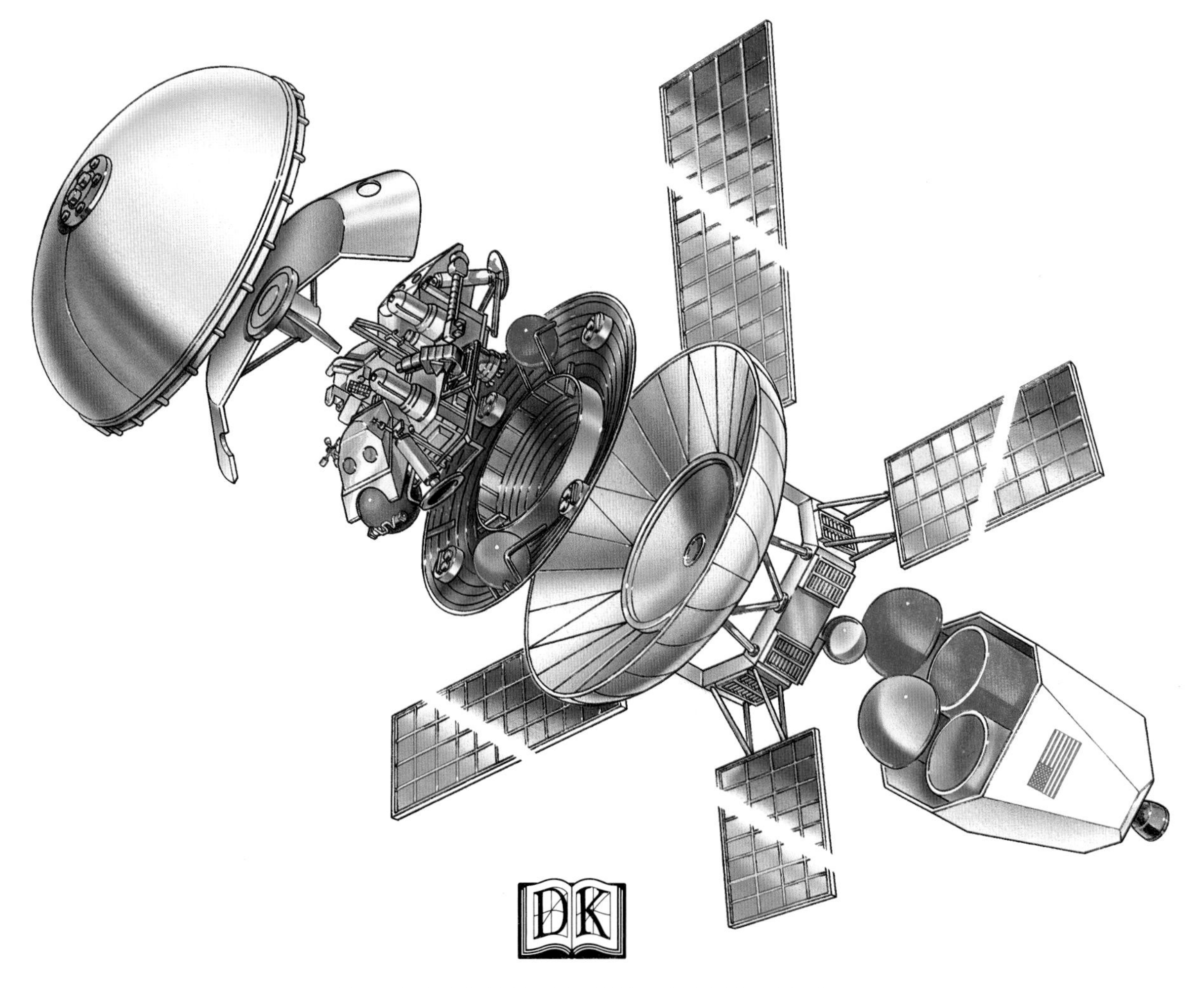

DK

DORLING KINDERSLEY

LONDON • NEW YORK • STUTTGART

A DORLING KINDERSLEY BOOK

Art Editor Dorian Spencer Davies
Designers Sharon Grant, Sara Hill
Senior Art Editor C. David Gillingwater
Project Editor Constance Novis
Senior Editor John C. Miles
Production Louise Barratt
Consultant Robin Kerrod

First published in 1994
by Dorling Kindersley Limited,
9 Henrietta Street, London WC2E 8PS

A CIP catalogue record for this book is available from the British Library

ISBN 0-7513-5218-7

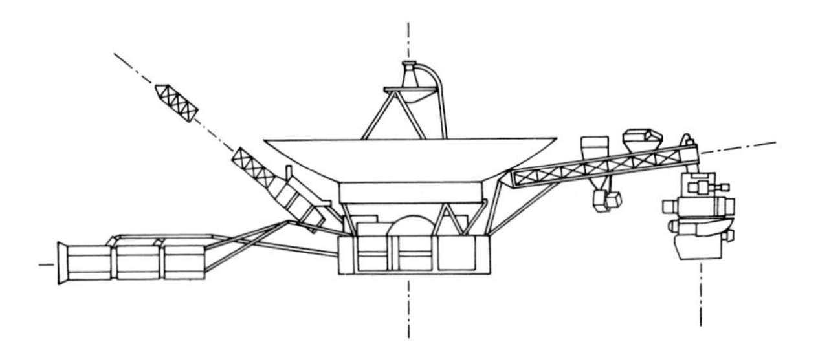

Reproduced by Dot Gradations, Essex
Printed and bound by Proost, Belgium

CONTENTS

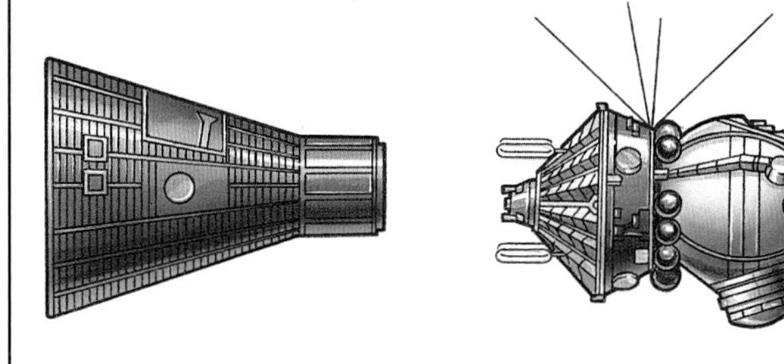

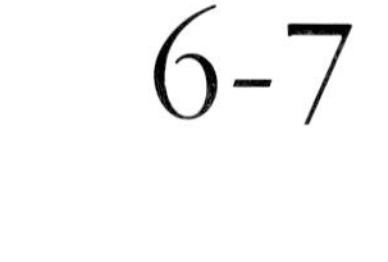

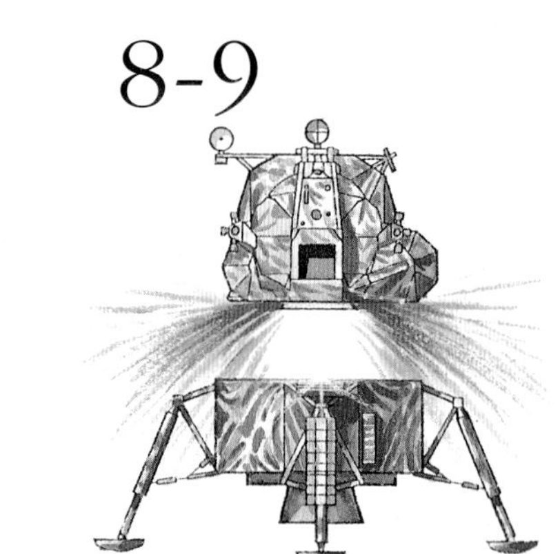

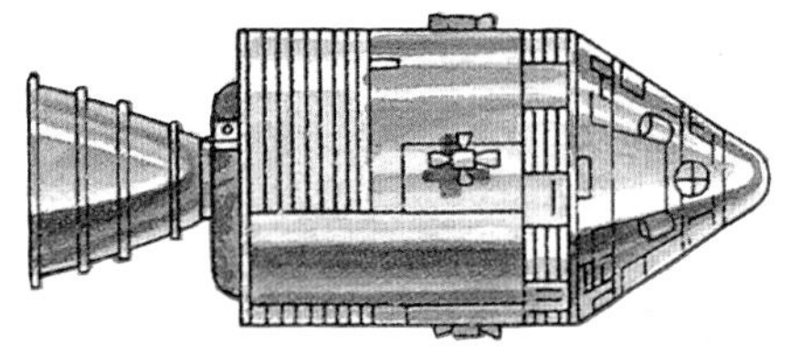

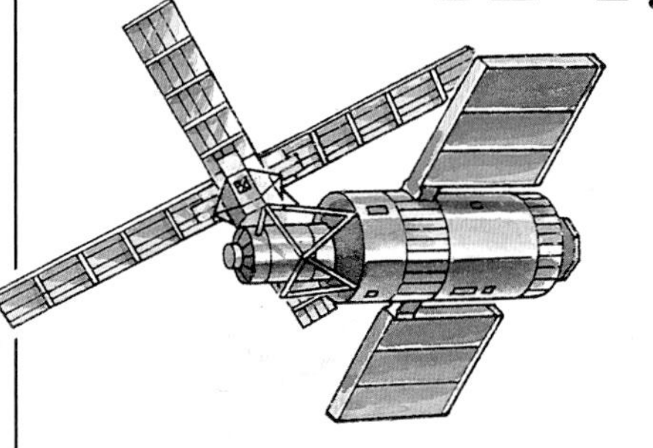

MERCURY

ON 4 OCTOBER 1957 THE USSR launched the world's first satellite, called Sputnik 1. As this small aluminium sphere hurtled through space it set off what came to be known as the "Space Race", with scientists of the USSR and the US competing to achieve supremacy in space. On 20 February 1962, the Americans put their first astronaut ("star sailor") into orbit in the Mercury spacecraft Friendship 7. His name was John Glenn and he became a national hero after he orbited the Earth three times in a trip lasting five hours.

Lift-off
A spaceship must blast off at high speed. Otherwise it would be pulled back by gravity. Friendship 7 was launched on top of a big rocket. Once the rocket had done its job it separated away from the manned capsule and fell back towards Earth.

Rescue rockets
The cone-shaped capsule had a rescue tower on top with an extra rocket. This could be used to separate the craft from the main rocket if something went wrong during the launch.

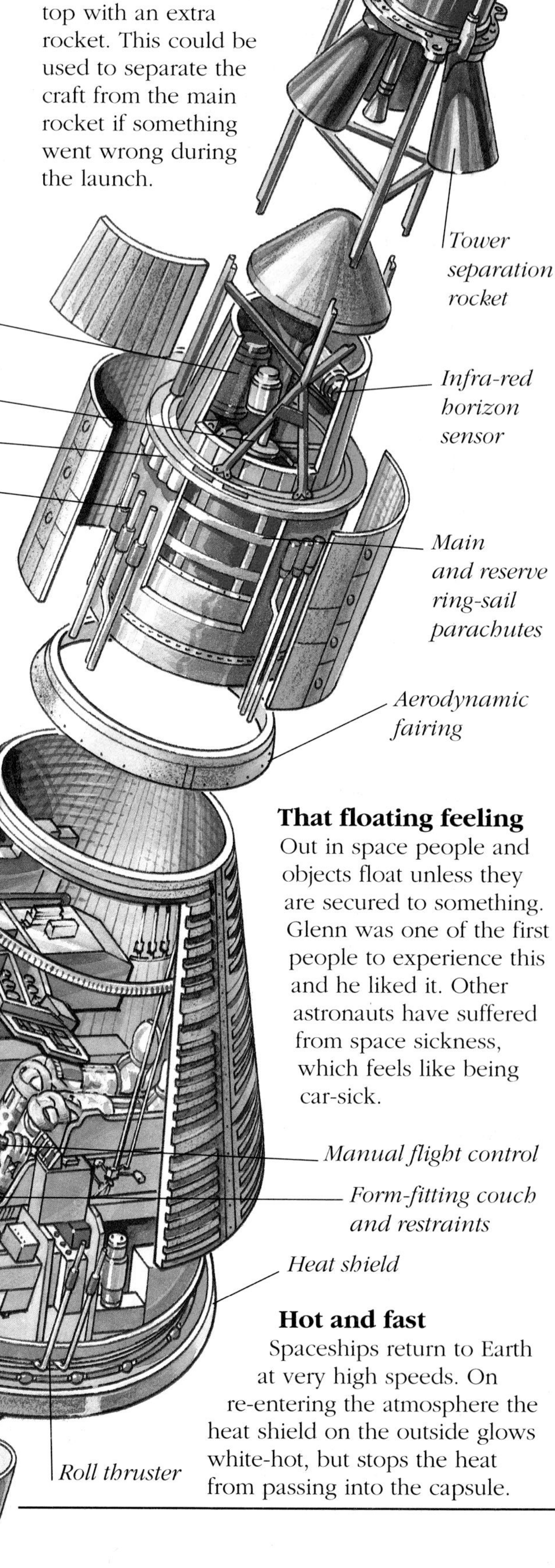

Under pressure
Around the Earth there is a layer of air called the atmosphere that pushes down on us. We need this pressure; without it our lungs wouldn't work. Out in space pressure has to be provided artificially, so John Glenn's capsule was pressurized using pure oxygen.

That floating feeling
Out in space people and objects float unless they are secured to something. Glenn was one of the first people to experience this and he liked it. Other astronauts have suffered from space sickness, which feels like being car-sick.

Hot and fast
Spaceships return to Earth at very high speeds. On re-entering the atmosphere the heat shield on the outside glows white-hot, but stops the heat from passing into the capsule.

TECHNICAL DATA

HEIGHT (INCLUDING TOWER): 7.9 M (26 FT)

HEIGHT (INCLUDING SPACECRAFT): 38.4 M (125 FT 10 IN)

DIAMETER: 23 M (90 FT 6 IN)

WIDTH ACROSS HEAT SHIELD: 1.89 M (6 FT 2 IN)

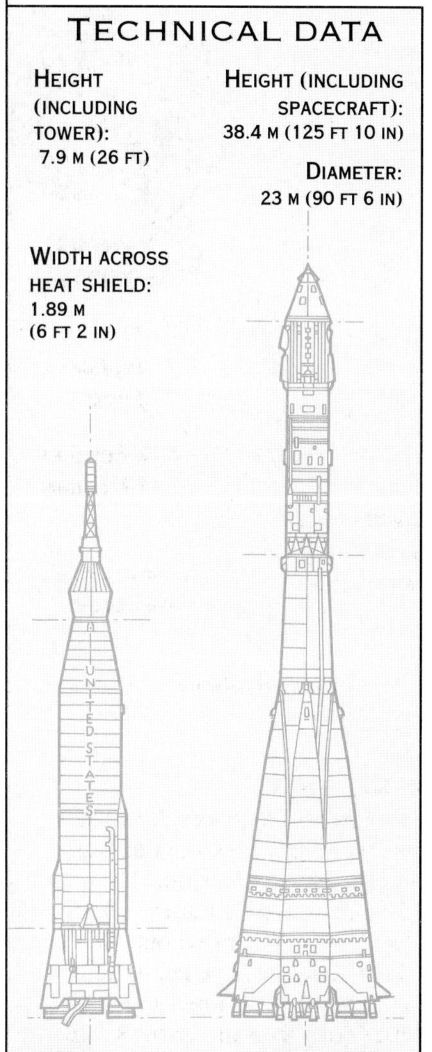

Voskhod 2

The USSR won the race to put a human into orbit on 12 April 1961 when they sent cosmonaut ("sailor of the cosmos") Yuri Gagarin into space in a Vostok spacecraft. Later versions of the craft were called Voskhods. This picture shows Voskhod 2, launched in 1965. One of its crew, Alexei Leonov, made the first ever walk in space on 18 March 1965.

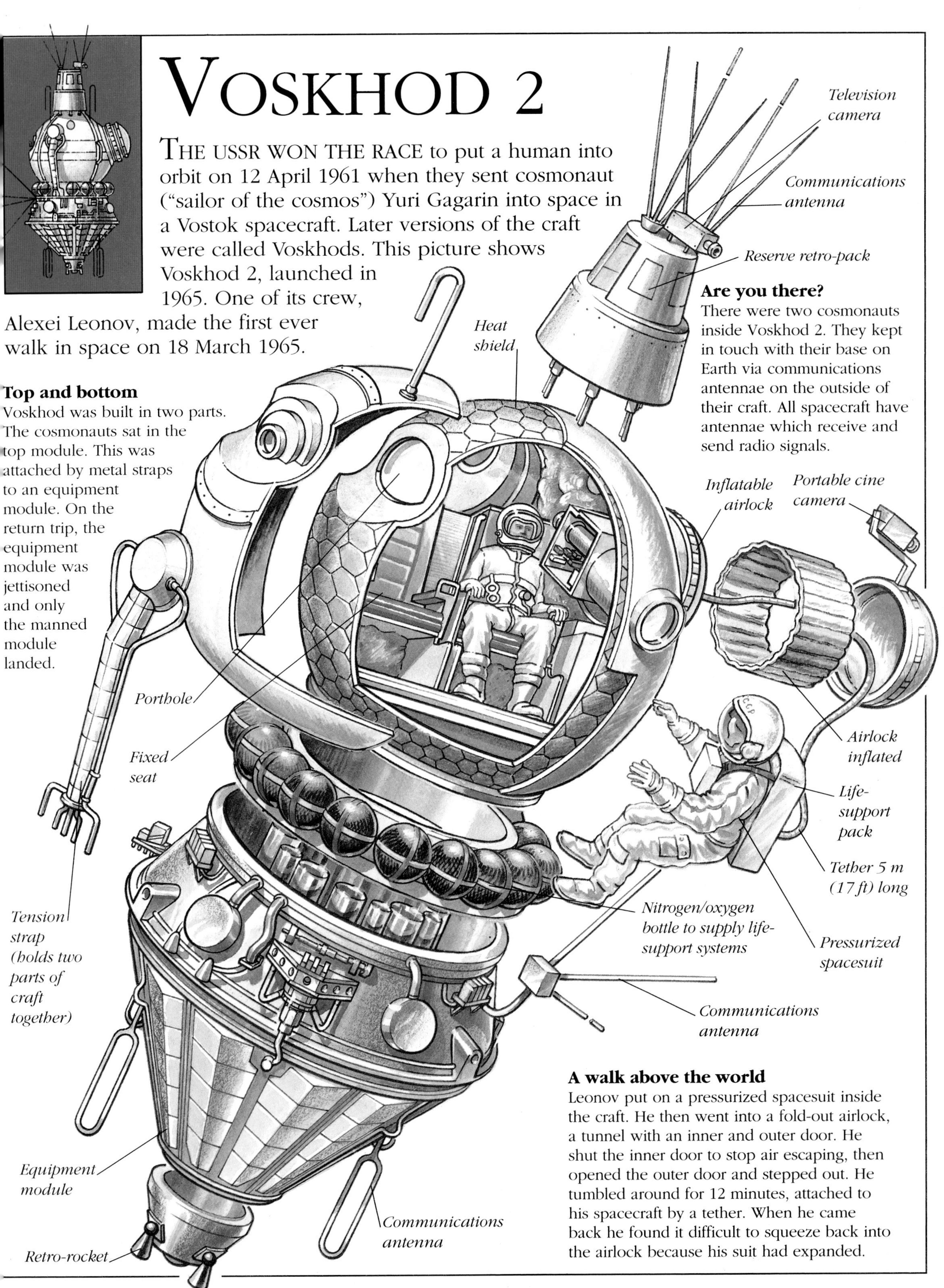

Top and bottom

Voskhod was built in two parts. The cosmonauts sat in the top module. This was attached by metal straps to an equipment module. On the return trip, the equipment module was jettisoned and only the manned module landed.

Are you there?

There were two cosmonauts inside Voskhod 2. They kept in touch with their base on Earth via communications antennae on the outside of their craft. All spacecraft have antennae which receive and send radio signals.

A walk above the world

Leonov put on a pressurized spacesuit inside the craft. He then went into a fold-out airlock, a tunnel with an inner and outer door. He shut the inner door to stop air escaping, then opened the outer door and stepped out. He tumbled around for 12 minutes, attached to his spacecraft by a tether. When he came back he found it difficult to squeeze back into the airlock because his suit had expanded.

SATURN V

"FIVE, FOUR, THREE, TWO, ONE... We have lift-off!" At the end of a launchpad countdown like this, the roar of giant engines fills the sky and the huge engine nozzles spit out columns of white-hot flame and gas, which push the entire rocket upwards. Between 1968 and 1972 giant Saturn V rockets boosted the American Apollo manned missions towards the Moon. The noise of a Saturn V launch sounded like a volcano erupting. A Saturn V took Apollo 11 into space on 16 July 1969. Four days later two of the crew members made history when they became the first humans ever to walk on the Moon.

Inside the engines
Inside a rocket there are separate tanks of liquid fuel and liquid oxygen, known as propellants. They are pumped into a combustion chamber inside each engine. There they are mixed and set alight to produce the hot gases needed to propel the rocket.

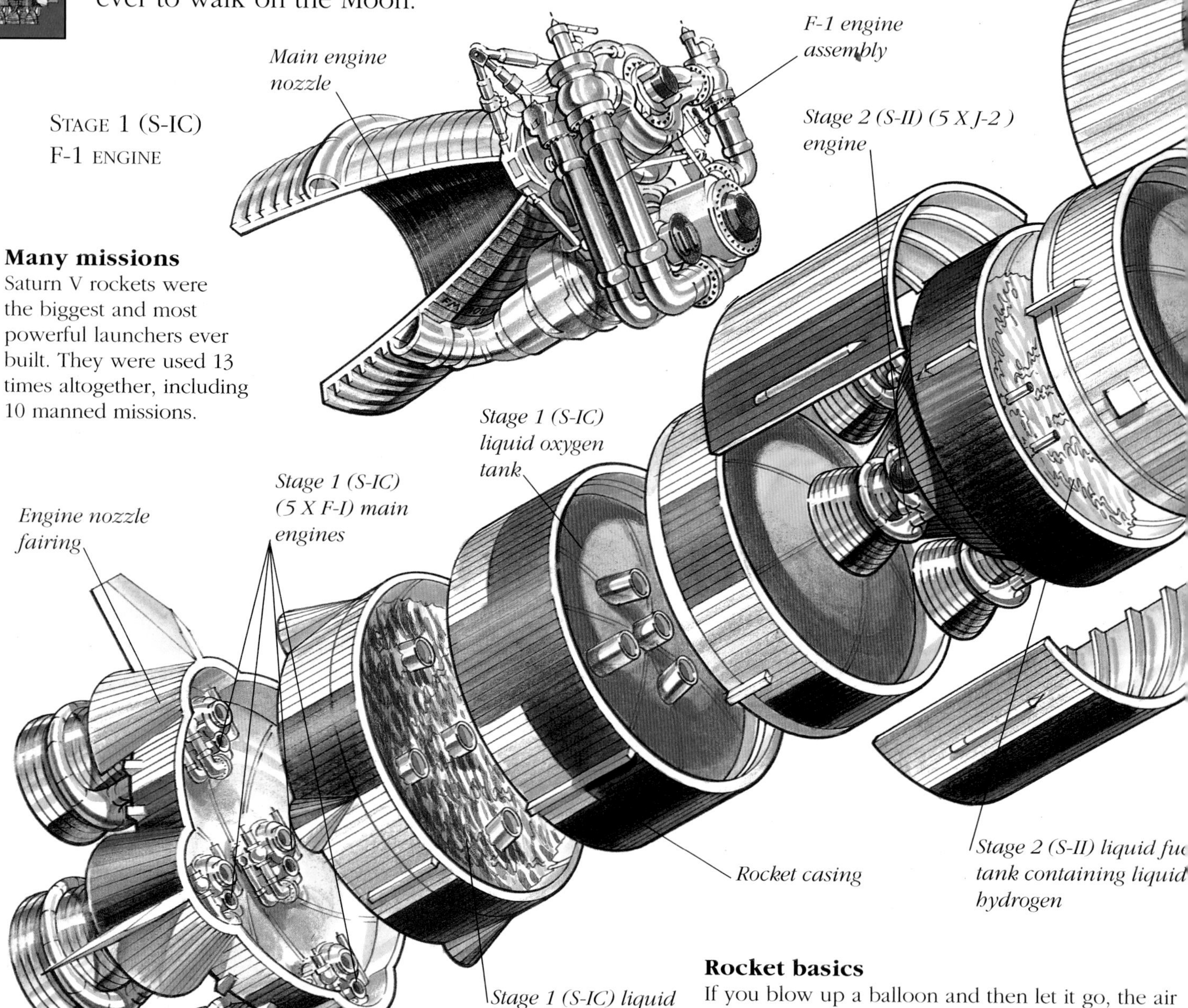

Many missions
Saturn V rockets were the biggest and most powerful launchers ever built. They were used 13 times altogether, including 10 manned missions.

Rocket basics
If you blow up a balloon and then let it go, the air rushes backwards out of the neck and pushes the balloon forwards. Rockets work in just the same way. Inside the rocket engines gases are made by burning fuel. The gases rush backwards out of the engine nozzles, pushing the rocket forwards.

Piece by piece

To produce enough power to escape Earth's gravity, the Saturn V needed three separate parts called stages. The first stage burnt for 2 min 30 sec, then separated and fell back to Earth. The second stage then fired and burnt for 6 min 30 sec. The third stage fired for about 2 min 30 sec to take Apollo into orbit round Earth. Then it fired again for about 5 min 30 sec to push Apollo towards the Moon before separating.

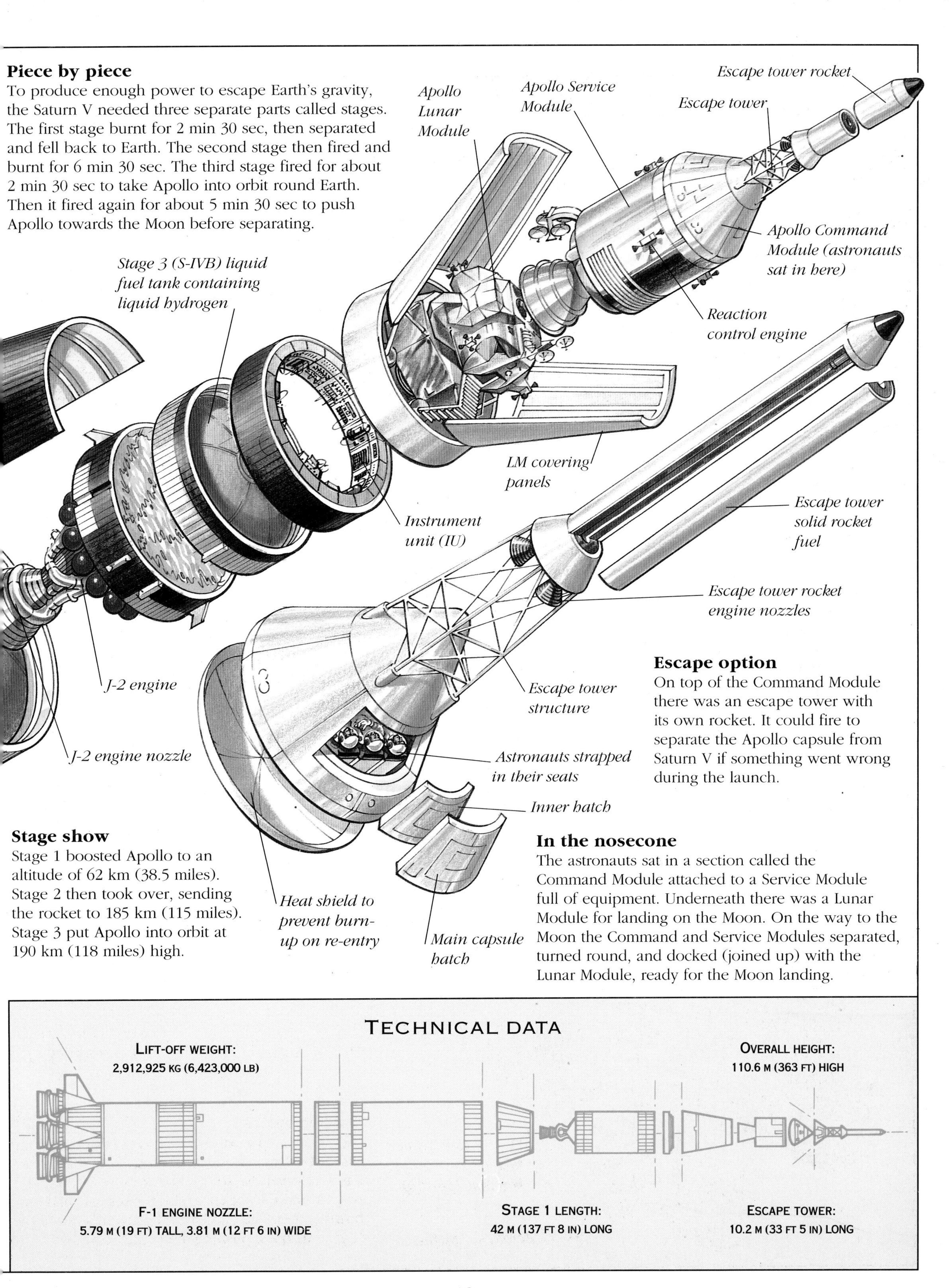

Escape option

On top of the Command Module there was an escape tower with its own rocket. It could fire to separate the Apollo capsule from Saturn V if something went wrong during the launch.

Stage show

Stage 1 boosted Apollo to an altitude of 62 km (38.5 miles). Stage 2 then took over, sending the rocket to 185 km (115 miles). Stage 3 put Apollo into orbit at 190 km (118 miles) high.

In the nosecone

The astronauts sat in a section called the Command Module attached to a Service Module full of equipment. Underneath there was a Lunar Module for landing on the Moon. On the way to the Moon the Command and Service Modules separated, turned round, and docked (joined up) with the Lunar Module, ready for the Moon landing.

Technical data

Lift-off weight: 2,912,925 kg (6,423,000 lb)

Overall height: 110.6 m (363 ft) high

F-1 engine nozzle: 5.79 m (19 ft) tall, 3.81 m (12 ft 6 in) wide

Stage 1 length: 42 m (137 ft 8 in) long

Escape tower: 10.2 m (33 ft 5 in) long

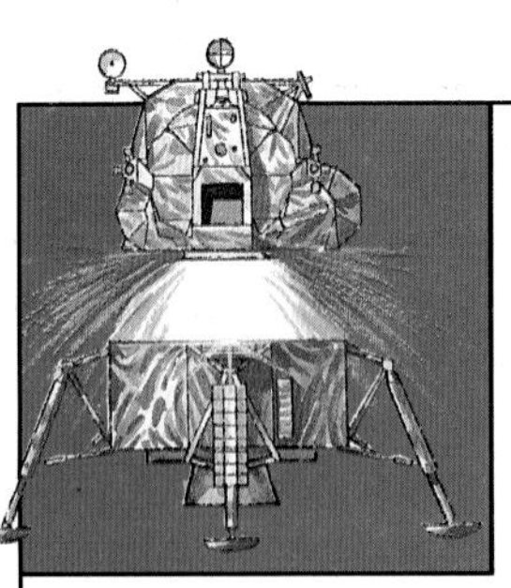

Apollo LM

ON 16 JULY 1969, A GIANT SATURN V rocket launched the world's most famous spacecraft – Apollo 11. NASA had launched a series of manned Apollo test flights which had gradually got closer to the Moon. Now the time had come to land! The Apollo astronauts went into orbit inside the Command Module. It docked with the Lunar Module and two crew members crawled through. The Lunar Module then separated and landed on the Moon. Part of it carried the crew back to the Command Module after their work was done. On 20 July 1969, Neil Armstrong left his Lunar Module, named "Eagle", and took the first step on the Moon's surface.

Space bug

The Lunar Module had a strange insect-like shape. Because the Lunar Module only operated in space where there is no air, its designers didn't need to worry about giving it a streamlined shape.

Reaction-control fuel

Reaction-control oxidizer

Docking hatch

VHF antenna

Relay box

Ascent fuel tank

Reaction-control pressurant

Ascent engine

Portable life-support system

Entry/Exit platform and rails

Rendezvous radar

Pilot's console

Entry hatch

Steerable antenna

Reaction-control thruster

Room inside

Inside the pressurized cabin there were computer consoles, viewing windows, and supplies. When the crew went out, they put on spacesuits and depressurized the cabin (so there was no air left in it). They opened a hatch, climbed out, and closed the door. When they came back they closed the door, repressurized the cabin, then took off their suits.

Taking off again

The "descent stage", the bottom section with the legs, served as a launchpad for the "ascent stage" when it was time to leave the Moon. It was left on the surface as the ascent stage flew up to dock with the Command and Service Modules. Once the crew were safely back in the Command Module, the ascent stage was jettisoned.

Landing

The Lunar Module lowered itself down on to the Moon. It had rockets to reduce its speed as it neared the surface, guided by radar. Its four fold-out legs had dishes on the bottom to support its weight. Three of them were fitted with sensing probes and as soon as these touched the surface they signalled to the crew to shut down the engines.

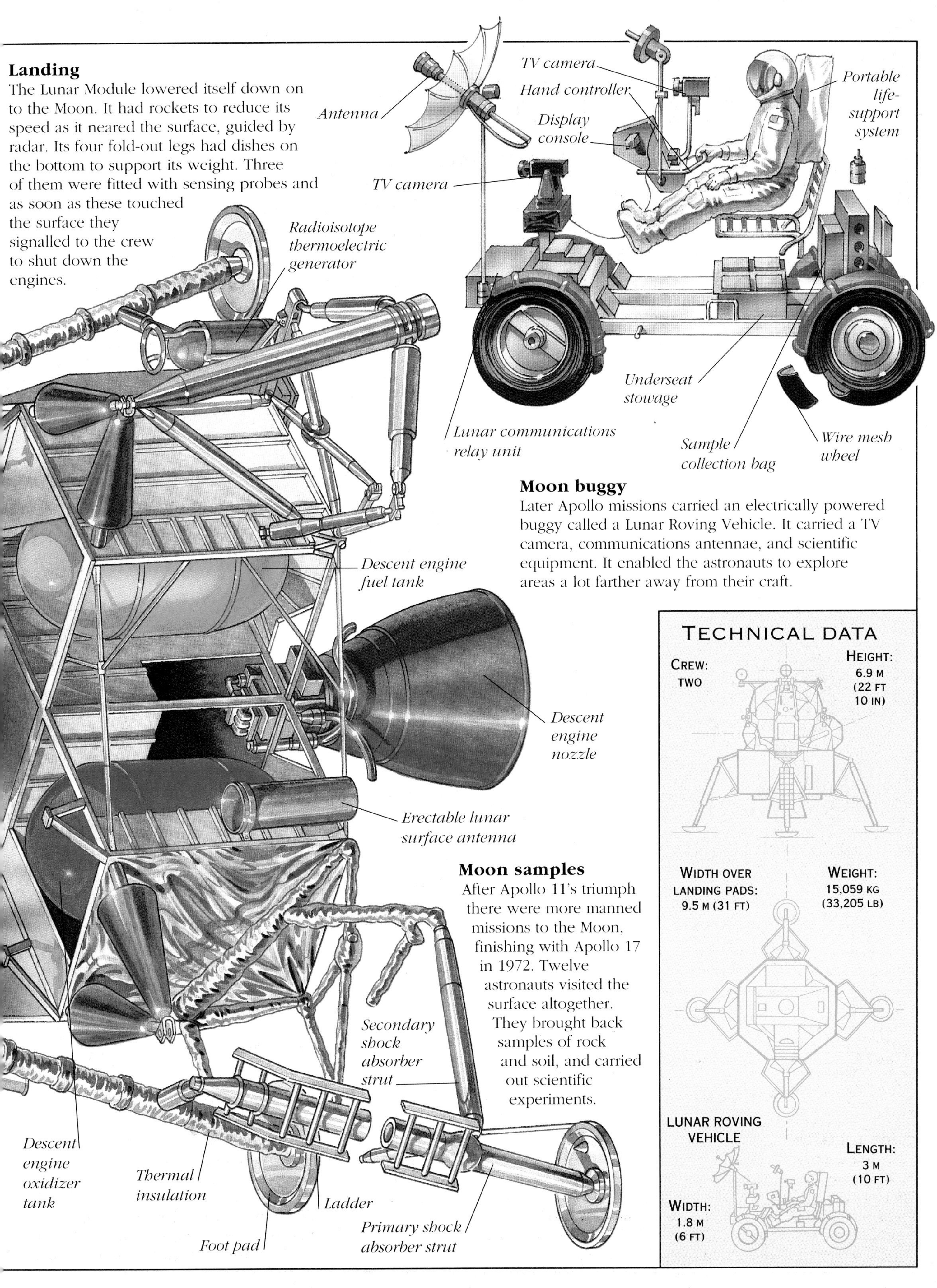

Moon buggy

Later Apollo missions carried an electrically powered buggy called a Lunar Roving Vehicle. It carried a TV camera, communications antennae, and scientific equipment. It enabled the astronauts to explore areas a lot farther away from their craft.

Moon samples

After Apollo 11's triumph there were more manned missions to the Moon, finishing with Apollo 17 in 1972. Twelve astronauts visited the surface altogether. They brought back samples of rock and soil, and carried out scientific experiments.

TECHNICAL DATA

CREW: TWO

HEIGHT: 6.9 M (22 FT 10 IN)

WIDTH OVER LANDING PADS: 9.5 M (31 FT)

WEIGHT: 15,059 KG (33,205 LB)

LUNAR ROVING VEHICLE

LENGTH: 3 M (10 FT)

WIDTH: 1.8 M (6 FT)

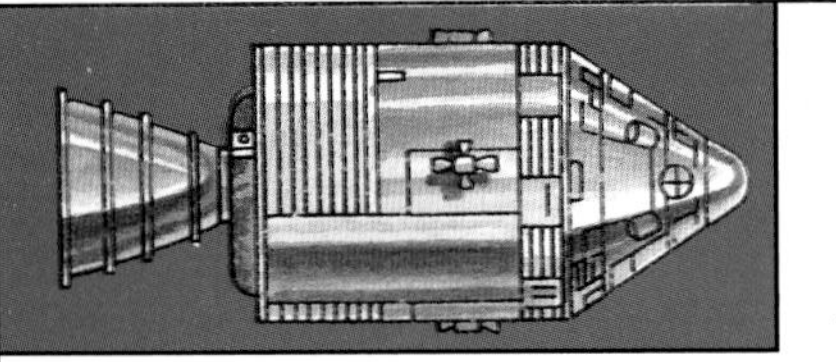

Apollo CSM

WHILE TWO APOLLO CREWMEN were busy on the lunar surface another stayed in the Command and Service Module (CSM for short), orbiting the Moon. When their work was finished the Moon-walking pair blasted off in the ascent stage of the Lunar Module. They docked with the CSM and crawled through a hatch into the Command Module cabin. The Lunar Module was jettisoned and then it was time to head for home.

Keeping in touch
NASA personnel at the Mission Control Center in Houston, Texas, kept in contact with the astronauts and monitored the workings of Apollo.

S-band antenna

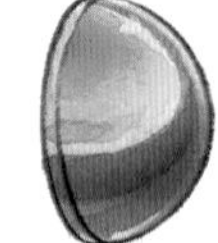

SERVICE MODULE

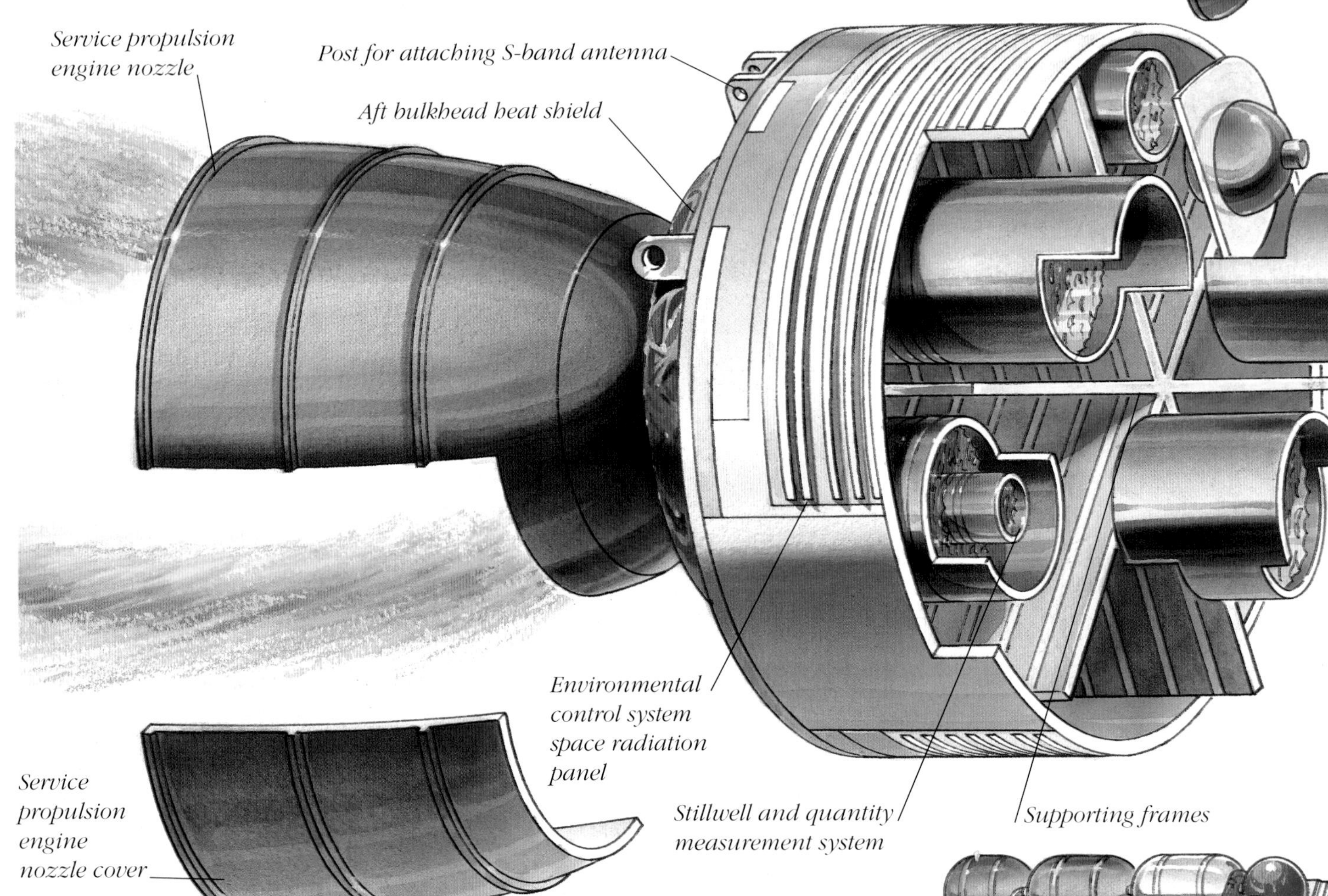

APOLLO 11 MISSION BADGE

Splashdown
The Command Module re-entered the Earth's atmosphere at about 39,000 km/h (24,200 mph). As it neared its journey's end parachutes opened to slow it down and it hit the water at 27 km/h (17 mph). Three airbags inflated like giant balloons to keep it upright in the water until US Navy helicopters could winch the crew to safety.

REACTION SYSTEM QUAD PANEL

Outer skin of SM

Propellant tank

Hello again!
The crew had radar to help them dock the CSM and the Lunar Module. The Command Module had a long probe that fitted into a dish-shape on the Lunar Module. The probe was guided into a hole in the centre of the dish and the two craft locked together. Then hatches were opened so that two of the crew could crawl through.

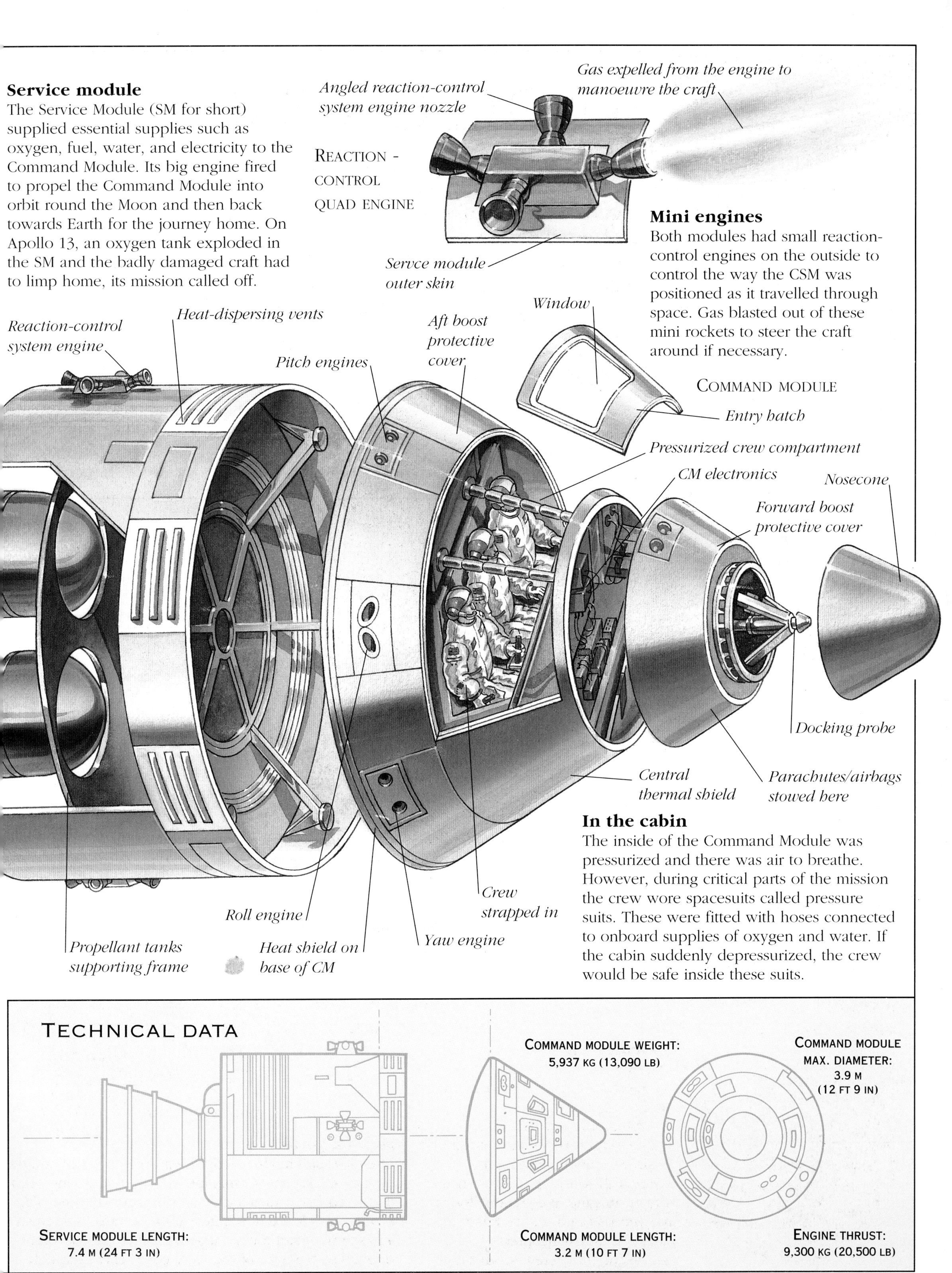

Service module

The Service Module (SM for short) supplied essential supplies such as oxygen, fuel, water, and electricity to the Command Module. Its big engine fired to propel the Command Module into orbit round the Moon and then back towards Earth for the journey home. On Apollo 13, an oxygen tank exploded in the SM and the badly damaged craft had to limp home, its mission called off.

Mini engines

Both modules had small reaction-control engines on the outside to control the way the CSM was positioned as it travelled through space. Gas blasted out of these mini rockets to steer the craft around if necessary.

In the cabin

The inside of the Command Module was pressurized and there was air to breathe. However, during critical parts of the mission the crew wore spacesuits called pressure suits. These were fitted with hoses connected to onboard supplies of oxygen and water. If the cabin suddenly depressurized, the crew would be safe inside these suits.

TECHNICAL DATA

COMMAND MODULE WEIGHT: 5,937 KG (13,090 LB)

COMMAND MODULE MAX. DIAMETER: 3.9 M (12 FT 9 IN)

SERVICE MODULE LENGTH: 7.4 M (24 FT 3 IN)

COMMAND MODULE LENGTH: 3.2 M (10 FT 7 IN)

ENGINE THRUST: 9,300 KG (20,500 LB)

Skylab

Once the US had succeeded In landing humans on the Moon, the next step was to build a space station where people could live and work. The Russians launched the first space station called Salyut in 1971. The Americans launched Skylab, shown below, in 1973. During 1973, and 1974, it was home to three different astronaut crews who travelled to it on board Apollo Command Modules. Their bodies were constantly monitored to see how well they coped with long-term life in space. They spent their time doing scientific experiments and taking photographs. They also had to repair their station. It was so badly damaged during its launch that it nearly didn't work at all!

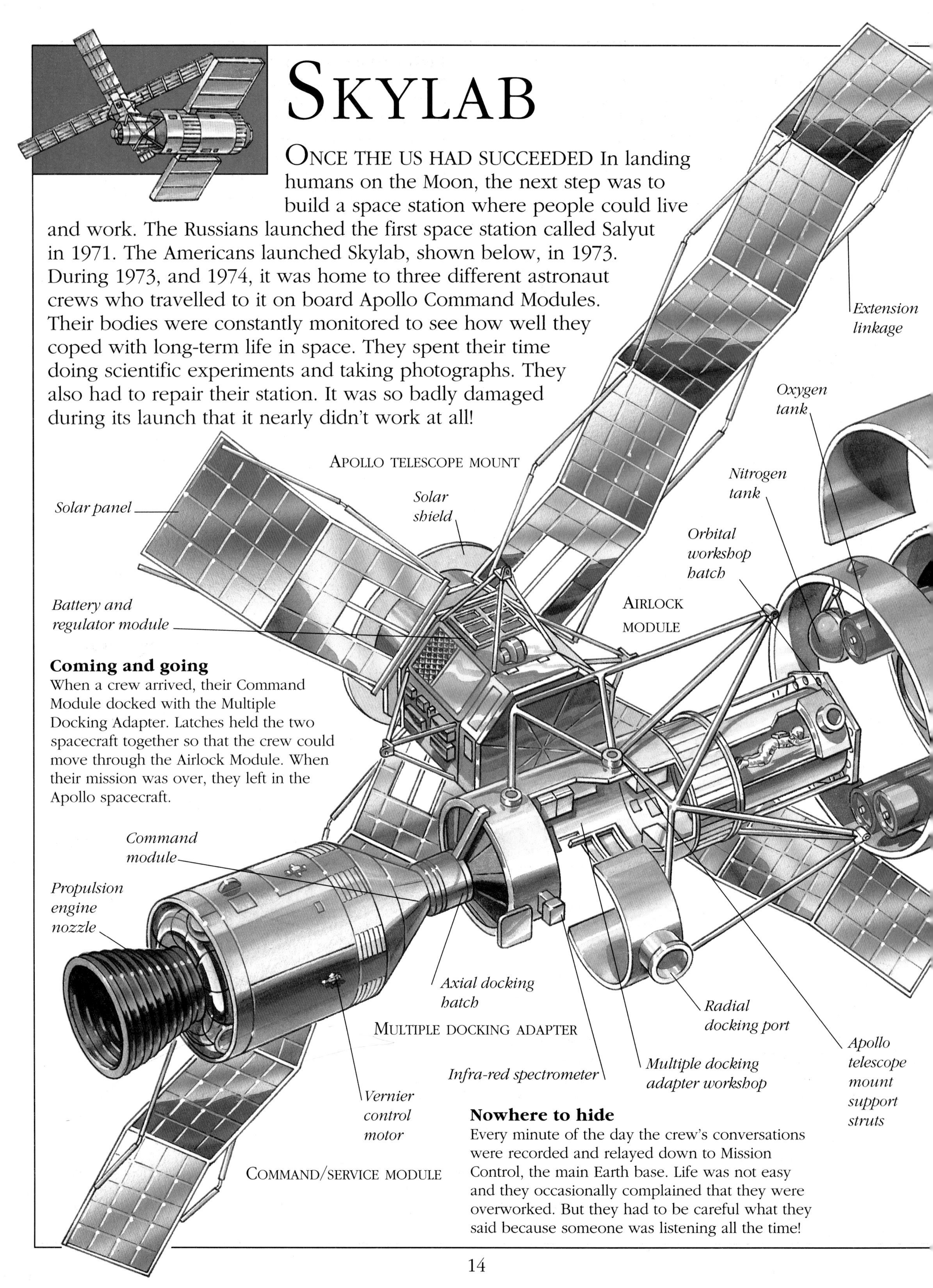

Coming and going

When a crew arrived, their Command Module docked with the Multiple Docking Adapter. Latches held the two spacecraft together so that the crew could move through the Airlock Module. When their mission was over, they left in the Apollo spacecraft.

Nowhere to hide

Every minute of the day the crew's conversations were recorded and relayed down to Mission Control, the main Earth base. Life was not easy and they occasionally complained that they were overworked. But they had to be careful what they said because someone was listening all the time!

Astronaut DIY

Skylab was fitted with huge solar arrays that converted the Sun's rays into electricity. During the launch these were badly damaged, leaving Skylab underpowered and overheated. However the first crew was able to repair them.

Living in a space home

Skylab was built using the third stage of a Saturn V rocket. One of its fuel tanks was converted into a workshop with a wardroom for relaxing in, a sleeping compartment, a bathroom, and an experiment room. Food and clothing were stored on board, including 210 pairs of underpants.

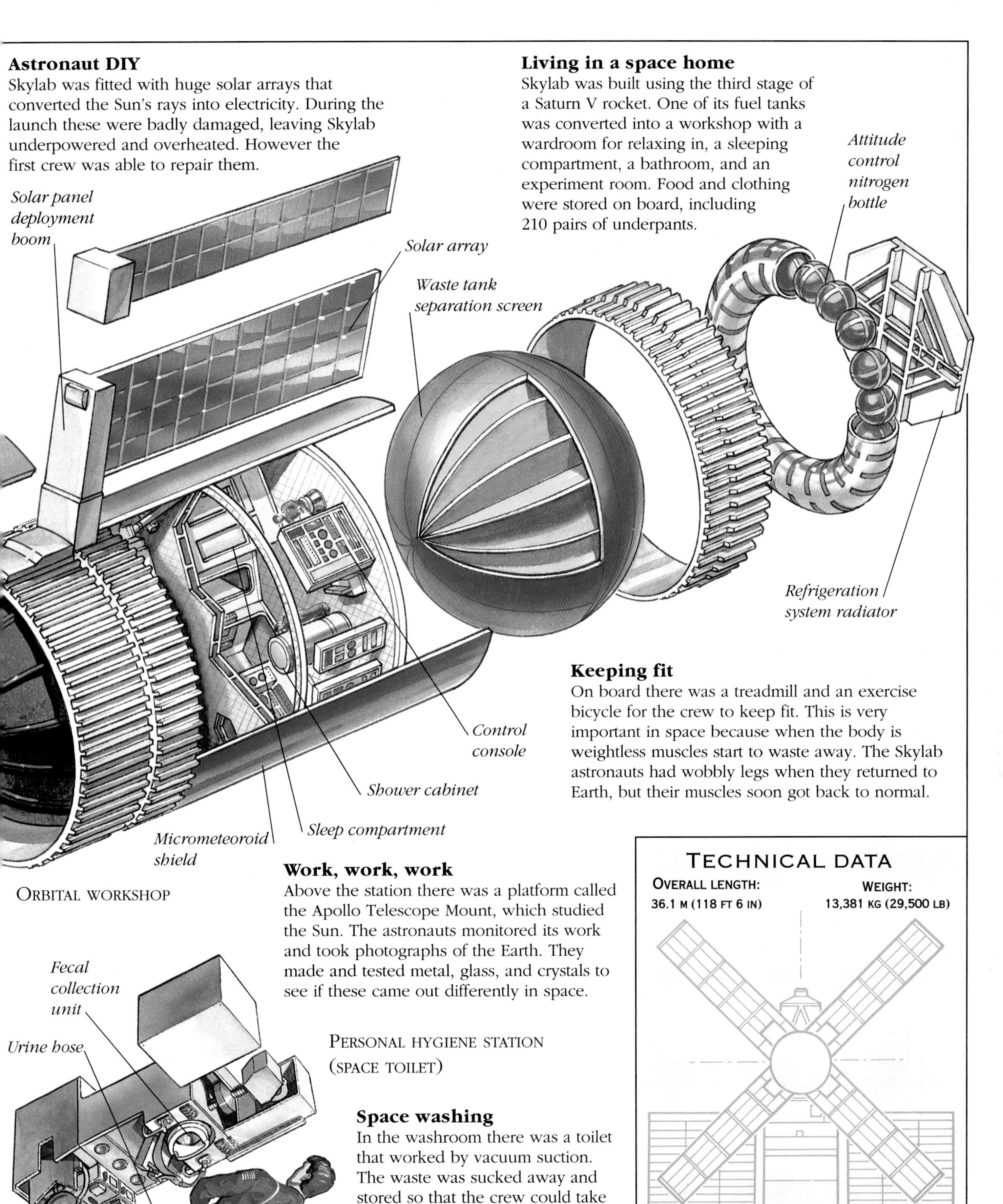

ORBITAL WORKSHOP

Keeping fit

On board there was a treadmill and an exercise bicycle for the crew to keep fit. This is very important in space because when the body is weightless muscles start to waste away. The Skylab astronauts had wobbly legs when they returned to Earth, but their muscles soon got back to normal.

Work, work, work

Above the station there was a platform called the Apollo Telescope Mount, which studied the Sun. The astronauts monitored its work and took photographs of the Earth. They made and tested metal, glass, and crystals to see if these came out differently in space.

PERSONAL HYGIENE STATION (SPACE TOILET)

Space washing

In the washroom there was a toilet that worked by vacuum suction. The waste was sucked away and stored so that the crew could take it back to Earth for study. An astronaut wanting a shower climbed into a collapsible tube device with a lid on top to stop water globules floating away.

TECHNICAL DATA

OVERALL LENGTH: 36.1 M (118 FT 6 IN)

WEIGHT: 13,381 KG (29,500 LB)

WORKING VOLUME: 331.5 M^3 (11,700 FT^3)

SOLAR PANEL LENGTH: 30 M (98 FT)

VIKING PROBE

UNMANNED SPACE PROBES are complex robot explorers, journeying far away to other planets. They send back fascinating new information and pictures. This is one of two Viking probes, both launched in 1975 towards the planet Mars. People have always wondered about outer space. Are there really Martians living on Mars? The Viking probes investigated puzzles like this by surveying the planet and searching for lifeforms.

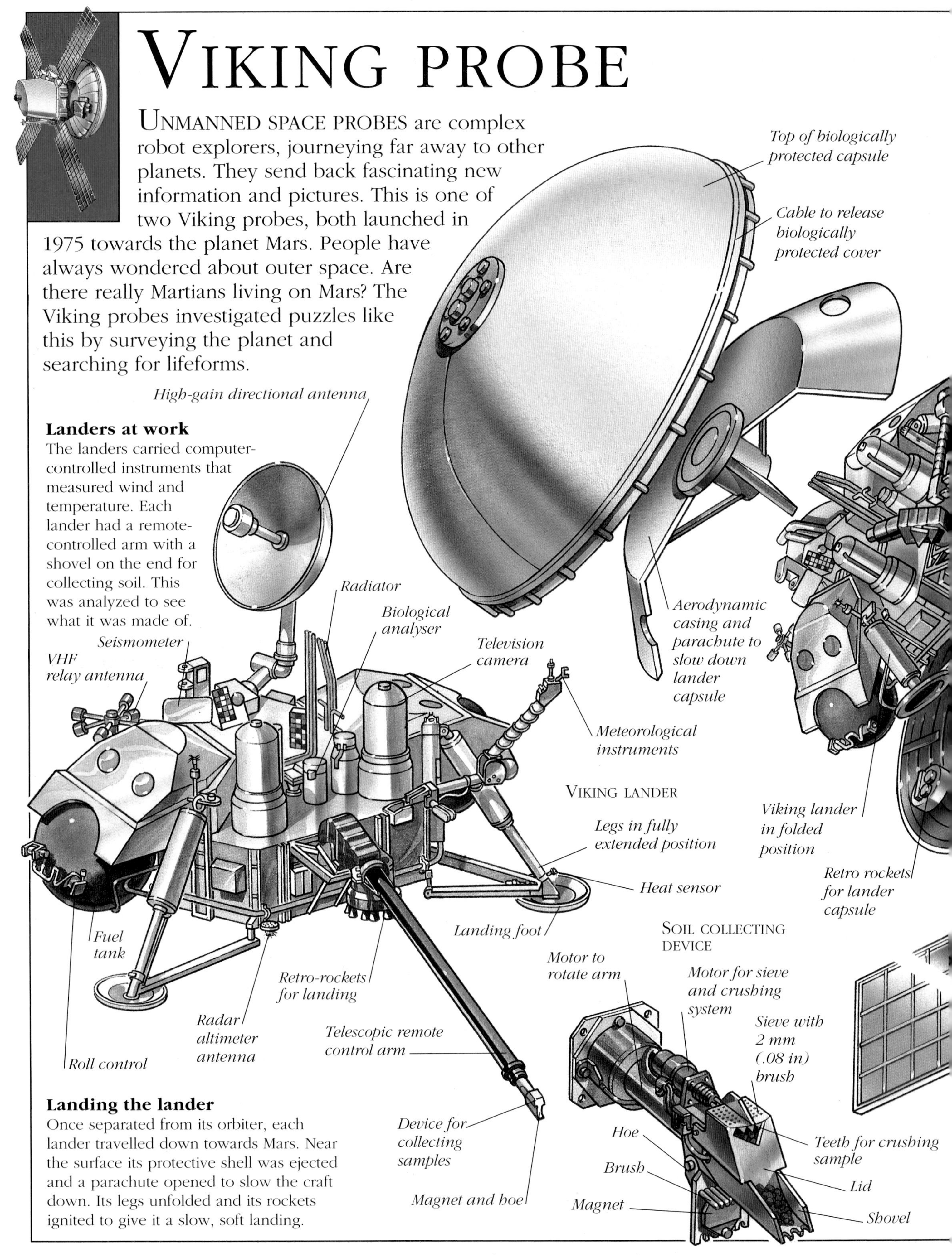

Landers at work

The landers carried computer-controlled instruments that measured wind and temperature. Each lander had a remote-controlled arm with a shovel on the end for collecting soil. This was analyzed to see what it was made of.

Landing the lander

Once separated from its orbiter, each lander travelled down towards Mars. Near the surface its protective shell was ejected and a parachute opened to slow the craft down. Its legs unfolded and its rockets ignited to give it a slow, soft landing.

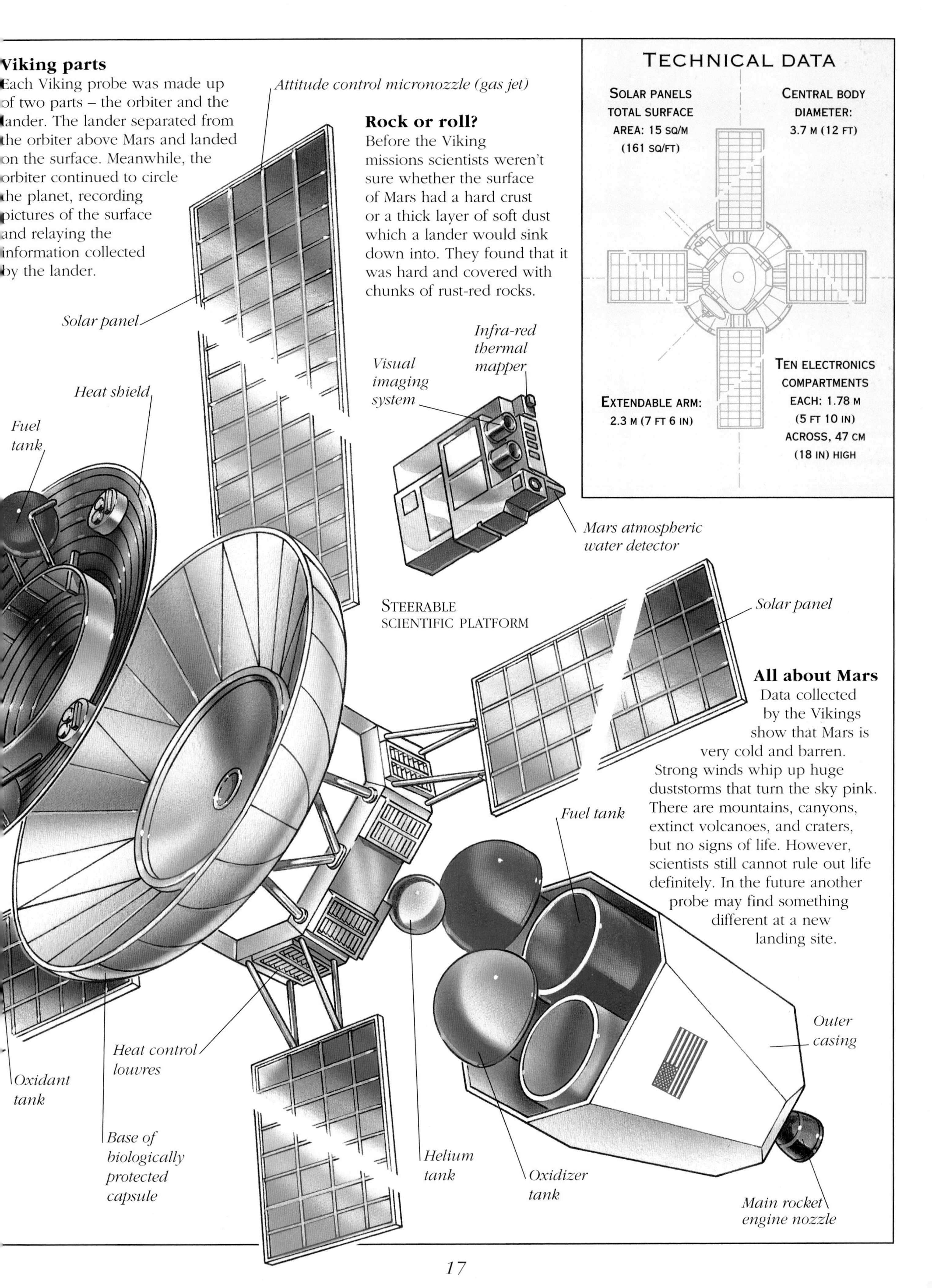

Viking parts
Each Viking probe was made up of two parts – the orbiter and the lander. The lander separated from the orbiter above Mars and landed on the surface. Meanwhile, the orbiter continued to circle the planet, recording pictures of the surface and relaying the information collected by the lander.

Rock or roll?
Before the Viking missions scientists weren't sure whether the surface of Mars had a hard crust or a thick layer of soft dust which a lander would sink down into. They found that it was hard and covered with chunks of rust-red rocks.

All about Mars
Data collected by the Vikings show that Mars is very cold and barren. Strong winds whip up huge duststorms that turn the sky pink. There are mountains, canyons, extinct volcanoes, and craters, but no signs of life. However, scientists still cannot rule out life definitely. In the future another probe may find something different at a new landing site.

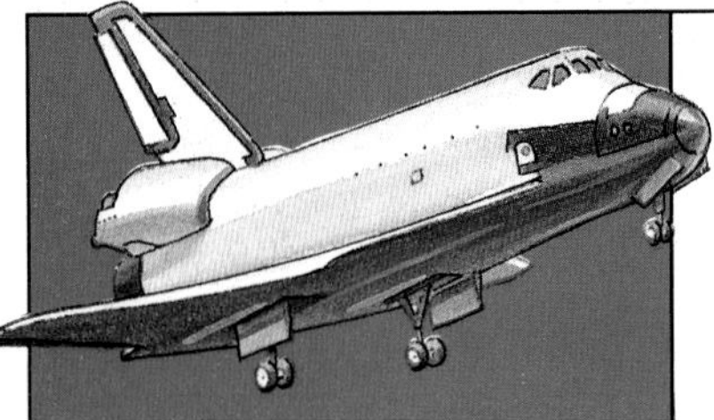

SPACE SHUTTLE

THE FIRST US SPACE SHUTTLE was launched on 12 April 1981. The Shuttle is reusable and since that first journey many more missions have been flown, teaching astronauts a lot more about living and working in space. The Shuttle Orbiter is a cross between a space station and a space plane. People can live inside it as it orbits the Earth. It can land its crew safely back home and then, after being checked and refitted, it can fly on another mission. It is mainly used to launch satellites and to rescue them for repair.

Keeping cool
Heat-absorbing liquid is pumped through pipes around the Orbiter to collect heat. Eventually the pipes pass through radiators inside the payload-bay doors. The doors are left open while in orbit to lose the extra heat into space.

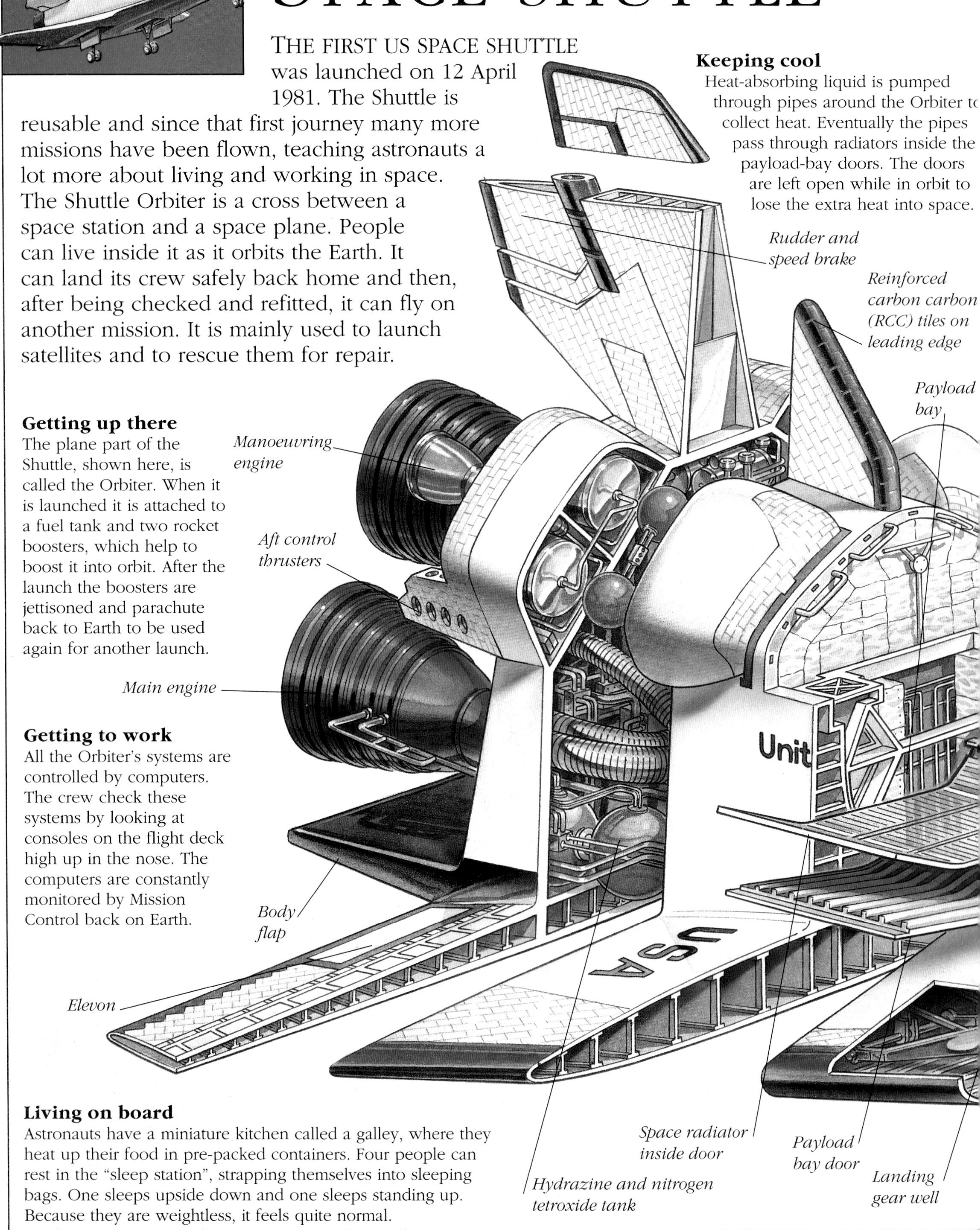

Getting up there
The plane part of the Shuttle, shown here, is called the Orbiter. When it is launched it is attached to a fuel tank and two rocket boosters, which help to boost it into orbit. After the launch the boosters are jettisoned and parachute back to Earth to be used again for another launch.

Getting to work
All the Orbiter's systems are controlled by computers. The crew check these systems by looking at consoles on the flight deck high up in the nose. The computers are constantly monitored by Mission Control back on Earth.

Living on board
Astronauts have a miniature kitchen called a galley, where they heat up their food in pre-packed containers. Four people can rest in the "sleep station", strapping themselves into sleeping bags. One sleeps upside down and one sleeps standing up. Because they are weightless, it feels quite normal.

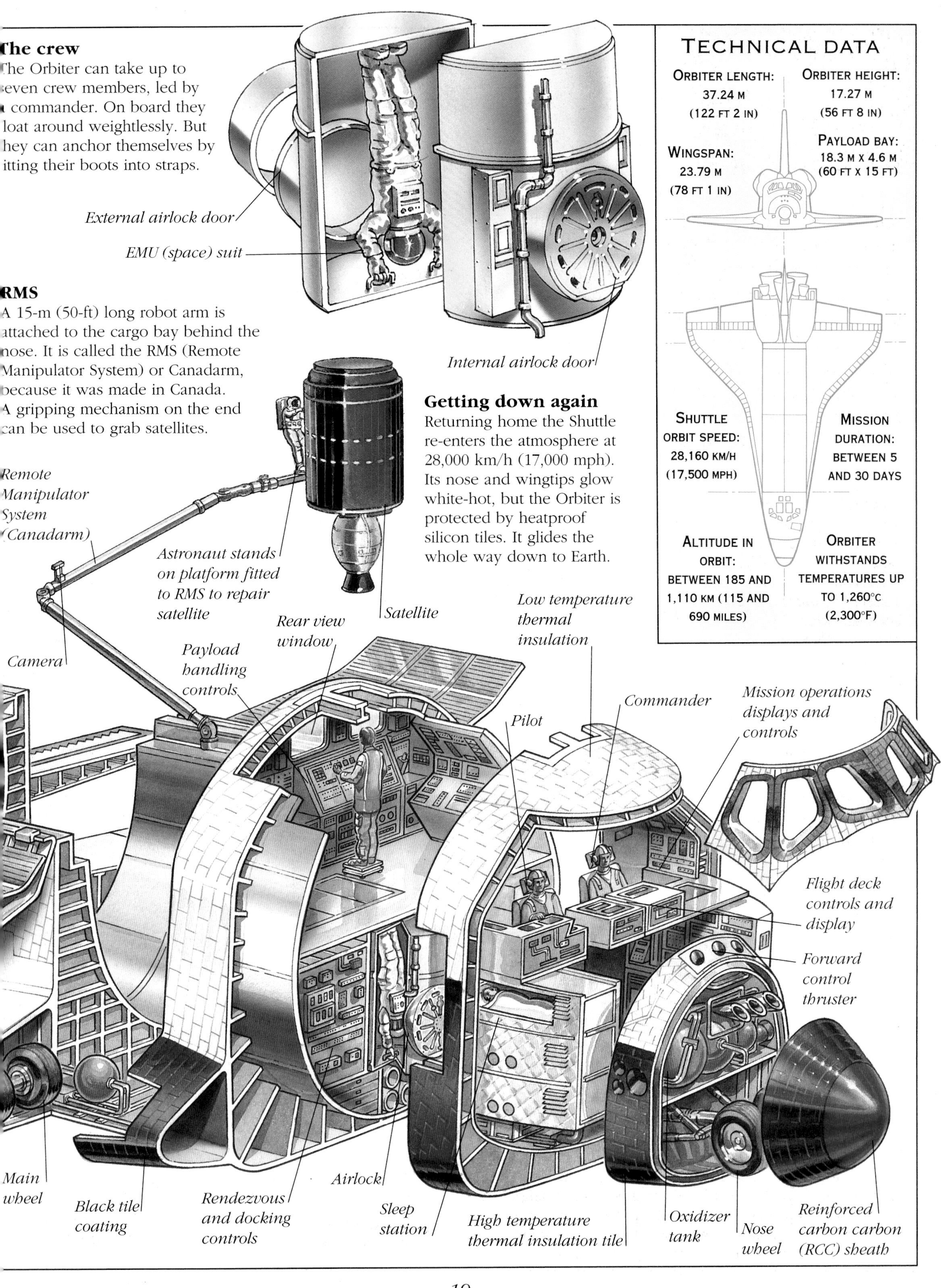

The crew

The Orbiter can take up to seven crew members, led by a commander. On board they float around weightlessly. But they can anchor themselves by fitting their boots into straps.

RMS

A 15-m (50-ft) long robot arm is attached to the cargo bay behind the nose. It is called the RMS (Remote Manipulator System) or Canadarm, because it was made in Canada. A gripping mechanism on the end can be used to grab satellites.

Getting down again

Returning home the Shuttle re-enters the atmosphere at 28,000 km/h (17,000 mph). Its nose and wingtips glow white-hot, but the Orbiter is protected by heatproof silicon tiles. It glides the whole way down to Earth.

TECHNICAL DATA

ORBITER LENGTH: 37.24 M (122 FT 2 IN)

ORBITER HEIGHT: 17.27 M (56 FT 8 IN)

WINGSPAN: 23.79 M (78 FT 1 IN)

PAYLOAD BAY: 18.3 M X 4.6 M (60 FT X 15 FT)

SHUTTLE ORBIT SPEED: 28,160 KM/H (17,500 MPH)

MISSION DURATION: BETWEEN 5 AND 30 DAYS

ALTITUDE IN ORBIT: BETWEEN 185 AND 1,110 KM (115 AND 690 MILES)

ORBITER WITHSTANDS TEMPERATURES UP TO 1,260°C (2,300°F)

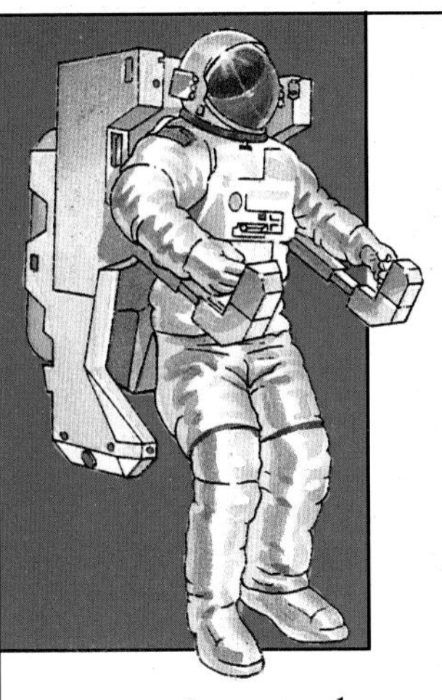

Space walk

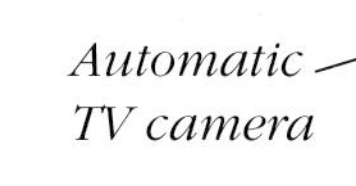

Sometimes astronauts have to go outside the Shuttle to capture and repair satellites or to test new equipment. But they would die a quick, painful death if they were not properly equipped. Going outside is called "extra-vehicular activity" or EVA for short. An astronaut on EVA must wear a complicated spacesuit for protection, to keep the correct pressure, and to provide air for breathing. Up until recently astronauts on EVA had to be tethered to their spacecraft by a cable. Otherwise they would float away and it would be very hard to rescue them. Now they can use a "Manned Manoeuvring Unit" or MMU for short. Shaped like the top part of an armchair, with arms and a back support, it jets the astronauts wherever they want to go.

Nitrogen gas tank

Strong metal ring locks suit parts together

Nitrogen gas tank holder

Technical data

MMU speed: up to 20 m/sec (66 ft/sec)

Cost of one suit: 3.6 million dollars

MMU weighs: 109 kg (240 lb) on Earth

MMU measurements: 1.25 m (5 ft) high, 0.83 m (2 ft 8 in) wide

Life-support backpack: 77 cm (50 in) high, 30 cm (19 in) deep, 58 cm (23 in) wide

MMU mission length: 7 hours max.

Astronaut underwear

Before a man puts on a spacesuit he attaches a urine collection device – a tube with a pouch on the end. Women wear layered shorts that absorb urine and draw it away. The next thing to put on is an undergarment fitted with tubes and holes called vents. These carry water to cool the astronaut.

Liquid cooling and vent undergarment

Tube carrying cooling water

Directional nitrogen gas nozzle

Outer suit trousers

Overshoe

Astronaut outerwear

The outer suit has parts that lock together with airtight metal rings. Made from many layers of nylon material, it has pleats in it that stretch to fit an astronaut's body. The inside of the suit is pumped with air to keep the astronaut's body pressurized.

Holes in socks keep astronaut' feet cool

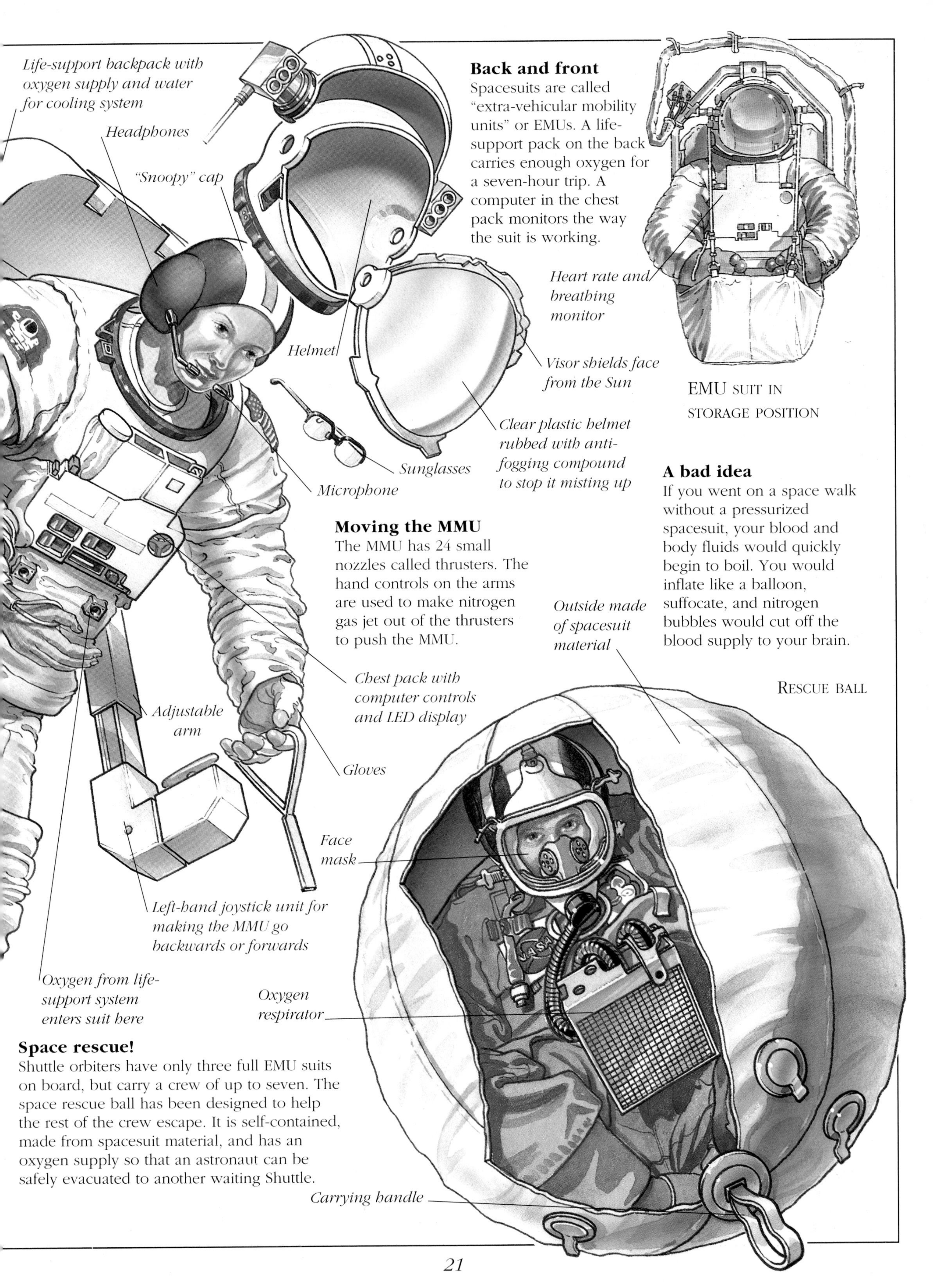

Back and front

Spacesuits are called "extra-vehicular mobility units" or EMUs. A life-support pack on the back carries enough oxygen for a seven-hour trip. A computer in the chest pack monitors the way the suit is working.

A bad idea

If you went on a space walk without a pressurized spacesuit, your blood and body fluids would quickly begin to boil. You would inflate like a balloon, suffocate, and nitrogen bubbles would cut off the blood supply to your brain.

Moving the MMU

The MMU has 24 small nozzles called thrusters. The hand controls on the arms are used to make nitrogen gas jet out of the thrusters to push the MMU.

Space rescue!

Shuttle orbiters have only three full EMU suits on board, but carry a crew of up to seven. The space rescue ball has been designed to help the rest of the crew escape. It is self-contained, made from spacesuit material, and has an oxygen supply so that an astronaut can be safely evacuated to another waiting Shuttle.

ARIANE 4

IN FRENCH GUIANA, SOUTH AMERICA, a massive launch site has been cut out of the jungle. This is the base for the European Ariane rockets, the workhorses of space. They regularly launch satellites into orbit round the Earth. They are "commercial space carriers", which means that any country can hire them to have satellites carried up and released. This picture shows the rocket most often used – the Ariane 4. It has three stages that separate from each other during flight, on the same principle as the Saturn V rocket on pages 8-9. It comes in six different versions which are chosen depending on how heavy the load is for each launch.

Command by computer
Ariane relies completely on its onboard computer. This commands the rocket stages to separate by triggering small explosive charges round the top of each stage. It also commands all the different engines to fire at the right times.

Booster rockets
The most powerful versions of Ariane 4 have extra booster rockets attached to stage 1 so they have the extra pushing power to launch heavy satellites. Some have two boosters; some have four.

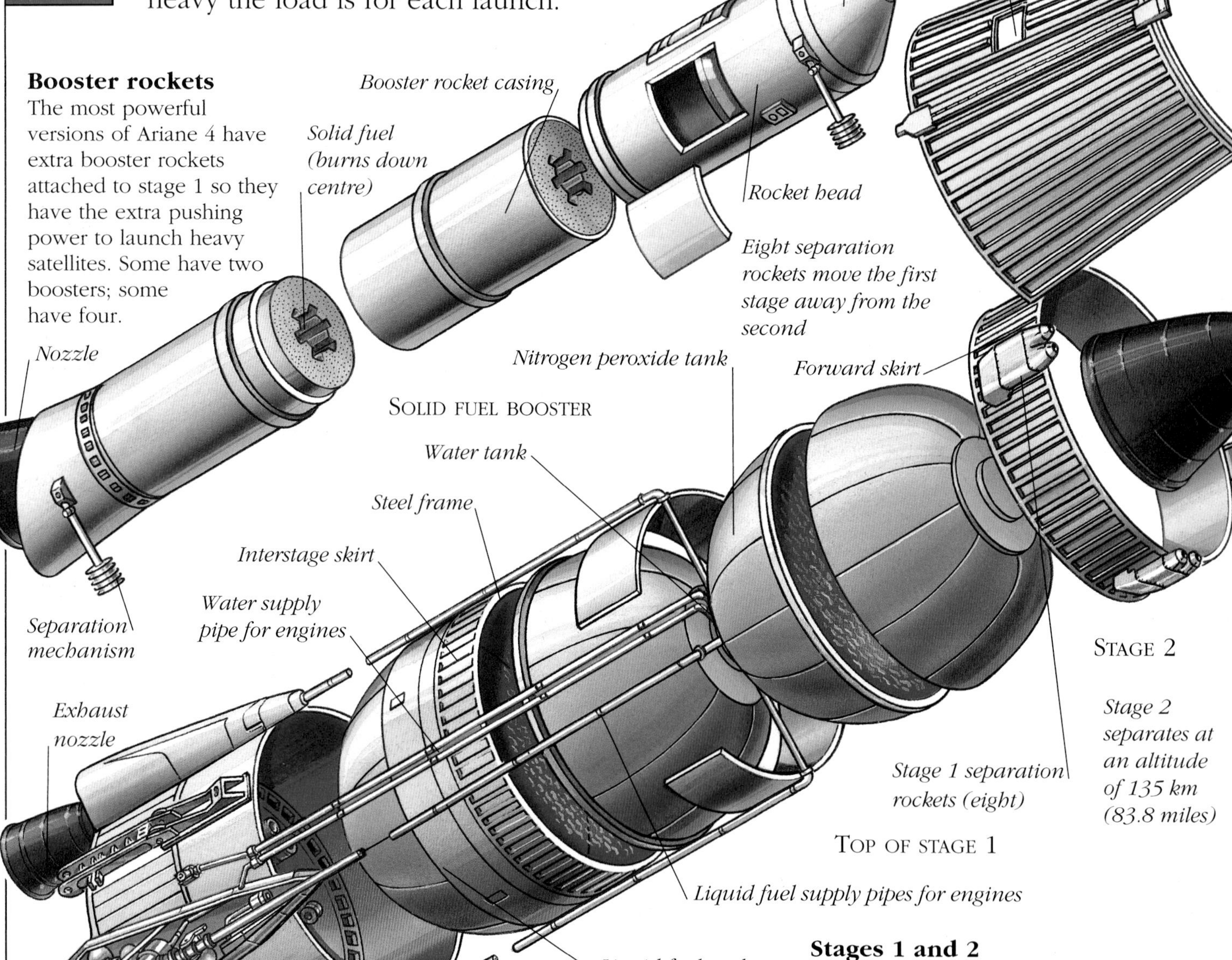

Stage 2 separates at an altitude of 135 km (83.8 miles)

Stages 1 and 2
Stage 1 and the booster rockets burn first. The boosters are jettisoned and then stage 1 separates. While the stage 2 engines are burning, the head is jettisoned, revealing the top satellite. Then stage 2 separates when its job is done.

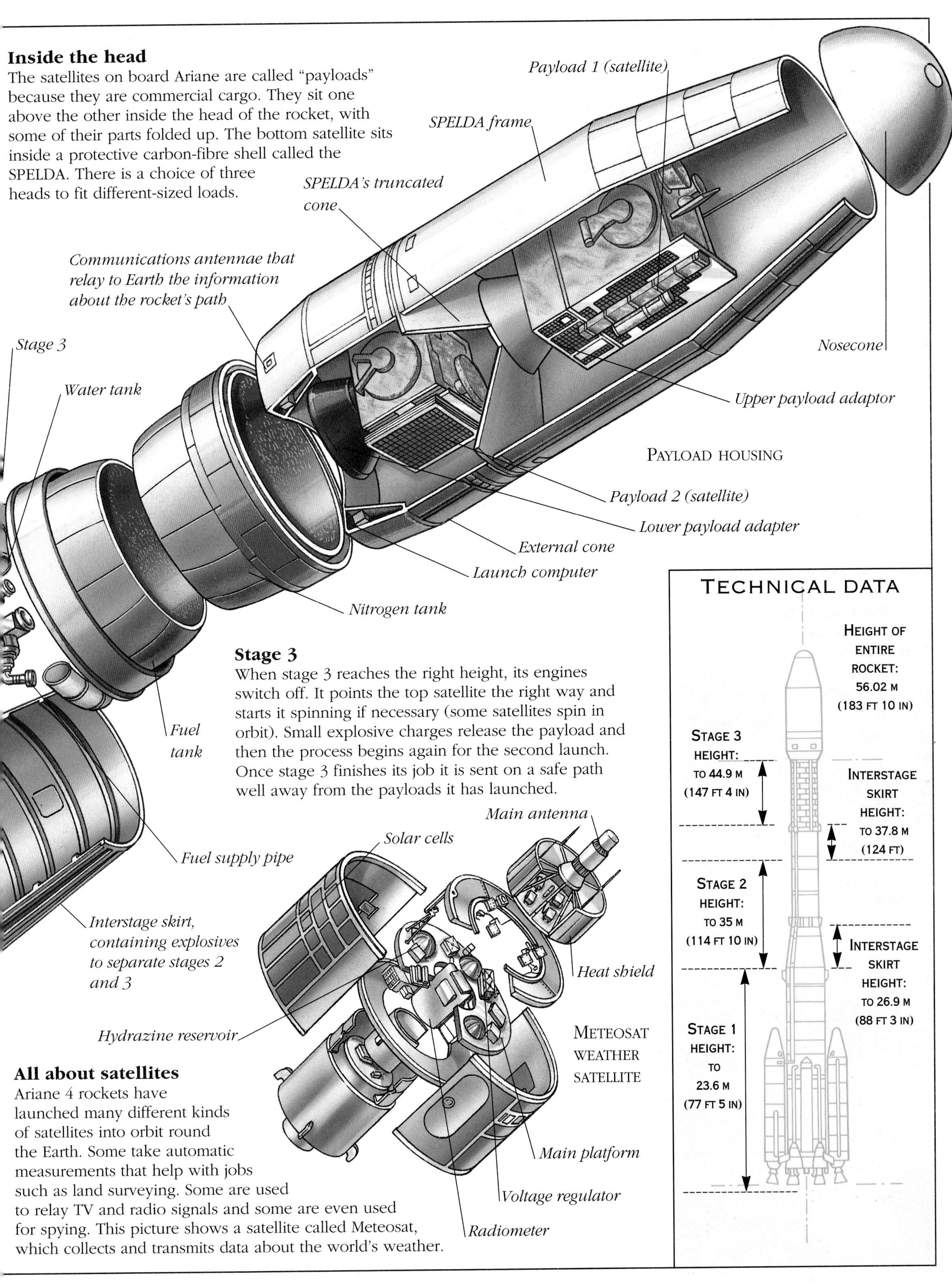

Inside the head

The satellites on board Ariane are called "payloads" because they are commercial cargo. They sit one above the other inside the head of the rocket, with some of their parts folded up. The bottom satellite sits inside a protective carbon-fibre shell called the SPELDA. There is a choice of three heads to fit different-sized loads.

Stage 3

When stage 3 reaches the right height, its engines switch off. It points the top satellite the right way and starts it spinning if necessary (some satellites spin in orbit). Small explosive charges release the payload and then the process begins again for the second launch. Once stage 3 finishes its job it is sent on a safe path well away from the payloads it has launched.

All about satellites

Ariane 4 rockets have launched many different kinds of satellites into orbit round the Earth. Some take automatic measurements that help with jobs such as land surveying. Some are used to relay TV and radio signals and some are even used for spying. This picture shows a satellite called Meteosat, which collects and transmits data about the world's weather.

VOYAGER

TWO VOYAGERS, 1 AND 2, were launched in 1977, beginning an exciting long-distance journey into deep space. They were sent to the outer area of the Solar System where they spied strange frozen moons and giant planets enveloped in poisonous gas. They are still travelling onwards through outer space, carrying a message from Earth to any intelligent beings.

On board the Voyagers

The Voyagers are controlled by computers. Their scientific instruments measure such things as magnetic fields, and their cameras send back spectacular images of the planets.

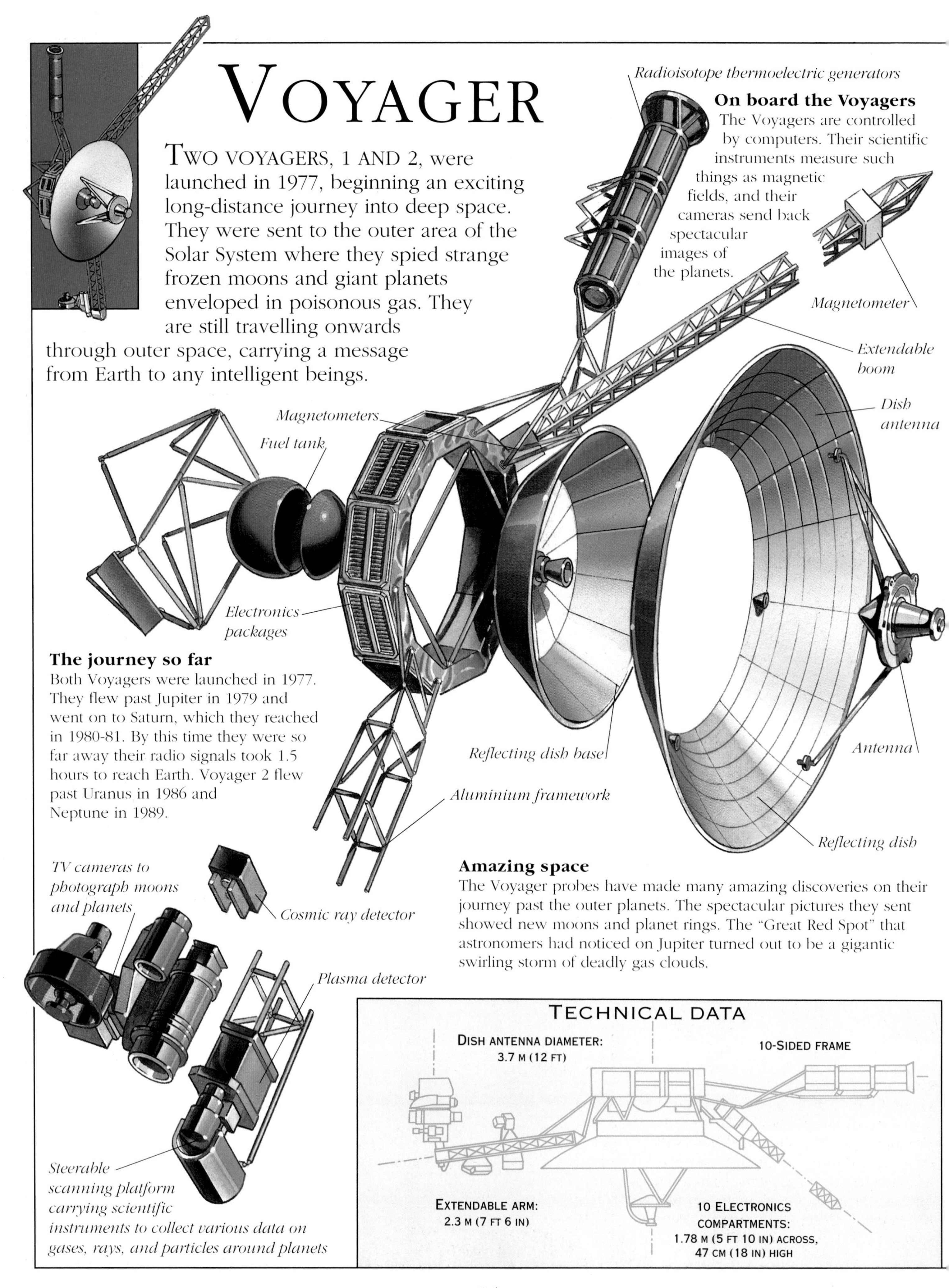

The journey so far

Both Voyagers were launched in 1977. They flew past Jupiter in 1979 and went on to Saturn, which they reached in 1980-81. By this time they were so far away their radio signals took 1.5 hours to reach Earth. Voyager 2 flew past Uranus in 1986 and Neptune in 1989.

Amazing space

The Voyager probes have made many amazing discoveries on their journey past the outer planets. The spectacular pictures they sent showed new moons and planet rings. The "Great Red Spot" that astronomers had noticed on Jupiter turned out to be a gigantic swirling storm of deadly gas clouds.

GIOTTO PROBE

GIOTTO WAS LAUNCHED IN 1985. In 1986 it passed close to Halley's Comet, taking measurements and pictures as it went. Halley's Comet is an object that orbits the Sun, passing the Earth every 76 years. For many centuries people thought it was a magical sign that heralded some great change. Scientists in modern times were keen to find out what it is made of.

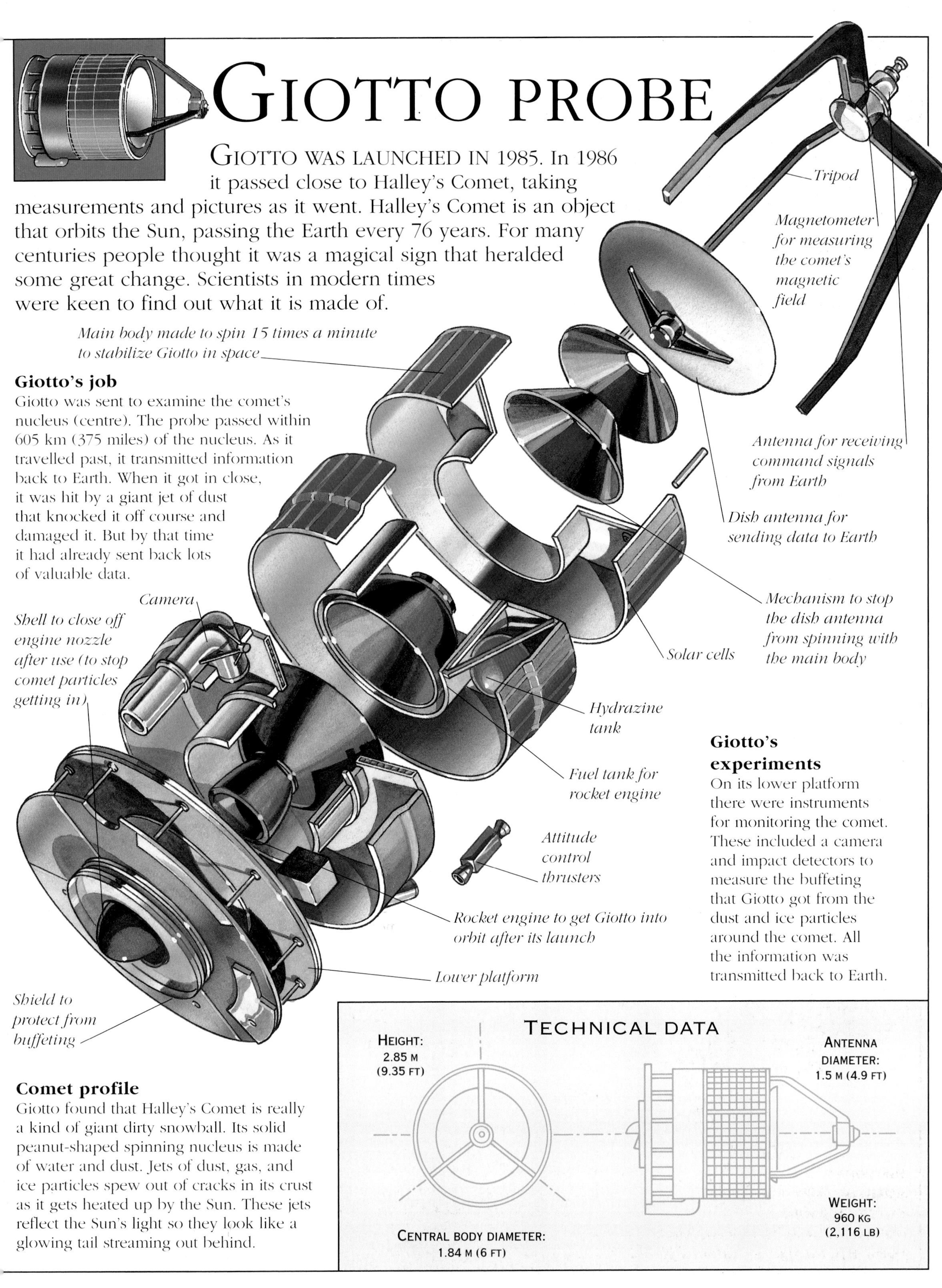

Giotto's job

Giotto was sent to examine the comet's nucleus (centre). The probe passed within 605 km (375 miles) of the nucleus. As it travelled past, it transmitted information back to Earth. When it got in close, it was hit by a giant jet of dust that knocked it off course and damaged it. But by that time it had already sent back lots of valuable data.

Giotto's experiments

On its lower platform there were instruments for monitoring the comet. These included a camera and impact detectors to measure the buffeting that Giotto got from the dust and ice particles around the comet. All the information was transmitted back to Earth.

Comet profile

Giotto found that Halley's Comet is really a kind of giant dirty snowball. Its solid peanut-shaped spinning nucleus is made of water and dust. Jets of dust, gas, and ice particles spew out of cracks in its crust as it gets heated up by the Sun. These jets reflect the Sun's light so they look like a glowing tail streaming out behind.

Hubble telescope

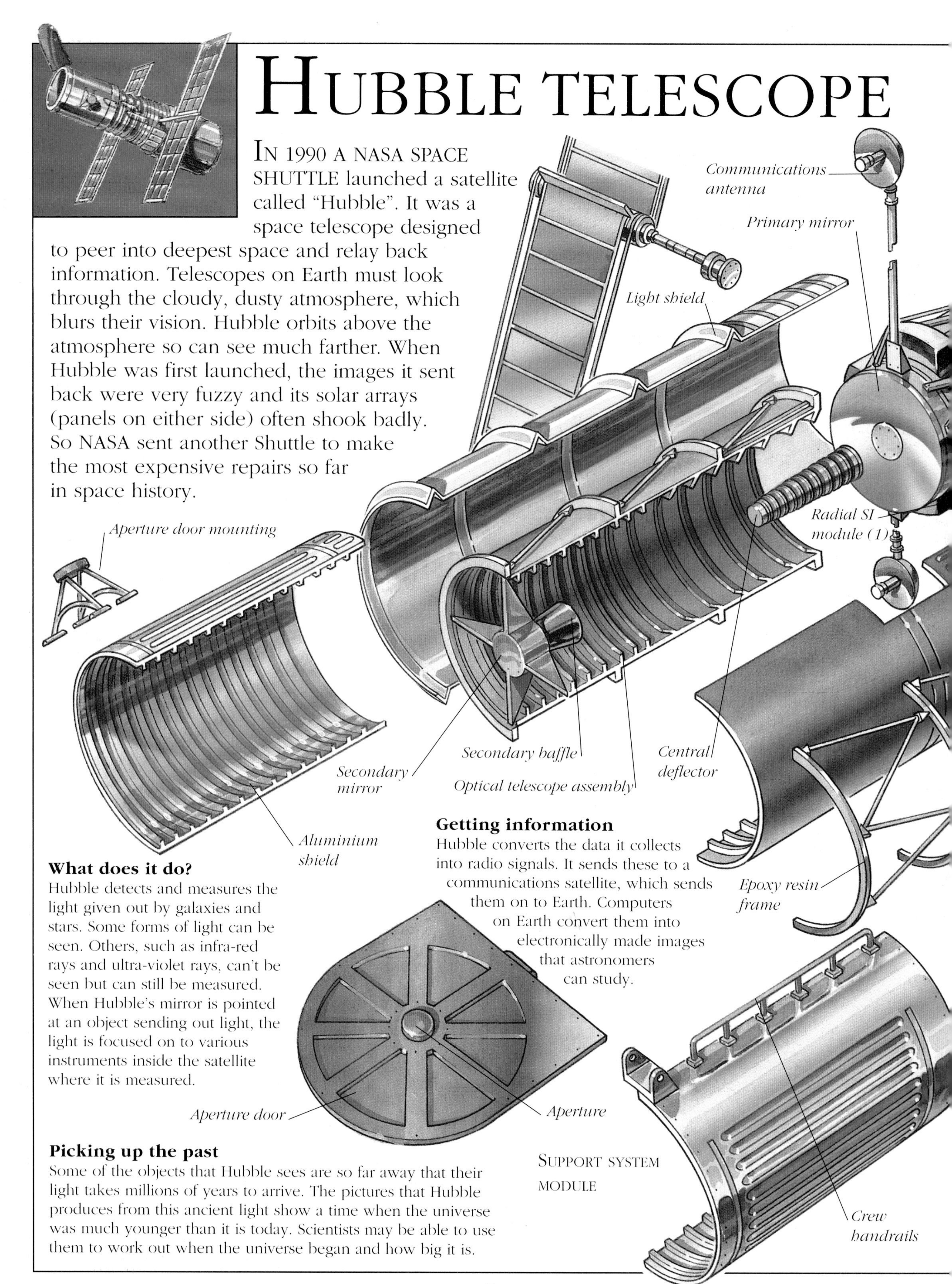

In 1990 a NASA space Shuttle launched a satellite called "Hubble". It was a space telescope designed to peer into deepest space and relay back information. Telescopes on Earth must look through the cloudy, dusty atmosphere, which blurs their vision. Hubble orbits above the atmosphere so can see much farther. When Hubble was first launched, the images it sent back were very fuzzy and its solar arrays (panels on either side) often shook badly. So NASA sent another Shuttle to make the most expensive repairs so far in space history.

What does it do?
Hubble detects and measures the light given out by galaxies and stars. Some forms of light can be seen. Others, such as infra-red rays and ultra-violet rays, can't be seen but can still be measured. When Hubble's mirror is pointed at an object sending out light, the light is focused on to various instruments inside the satellite where it is measured.

Getting information
Hubble converts the data it collects into radio signals. It sends these to a communications satellite, which sends them on to Earth. Computers on Earth convert them into electronically made images that astronomers can study.

Support system module

Picking up the past
Some of the objects that Hubble sees are so far away that their light takes millions of years to arrive. The pictures that Hubble produces from this ancient light show a time when the universe was much younger than it is today. Scientists may be able to use them to work out when the universe began and how big it is.

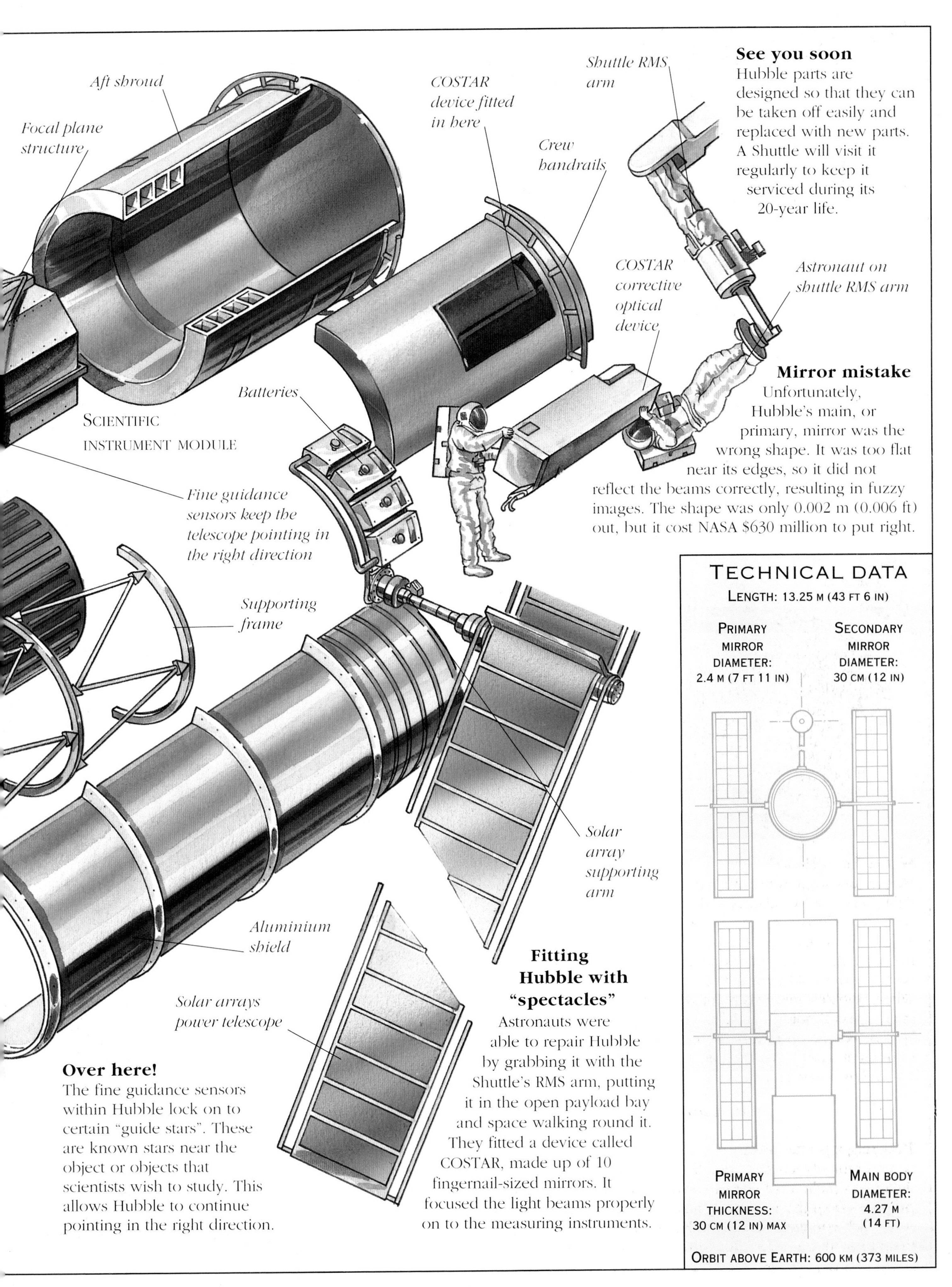

See you soon

Hubble parts are designed so that they can be taken off easily and replaced with new parts. A Shuttle will visit it regularly to keep it serviced during its 20-year life.

Mirror mistake

Unfortunately, Hubble's main, or primary, mirror was the wrong shape. It was too flat near its edges, so it did not reflect the beams correctly, resulting in fuzzy images. The shape was only 0.002 m (0.006 ft) out, but it cost NASA $630 million to put right.

Over here!

The fine guidance sensors within Hubble lock on to certain "guide stars". These are known stars near the object or objects that scientists wish to study. This allows Hubble to continue pointing in the right direction.

Fitting Hubble with "spectacles"

Astronauts were able to repair Hubble by grabbing it with the Shuttle's RMS arm, putting it in the open payload bay and space walking round it. They fitted a device called COSTAR, made up of 10 fingernail-sized mirrors. It focused the light beams properly on to the measuring instruments.

TECHNICAL DATA

LENGTH: 13.25 M (43 FT 6 IN)

PRIMARY MIRROR DIAMETER: 2.4 M (7 FT 11 IN)

SECONDARY MIRROR DIAMETER: 30 CM (12 IN)

PRIMARY MIRROR THICKNESS: 30 CM (12 IN) MAX

MAIN BODY DIAMETER: 4.27 M (14 FT)

ORBIT ABOVE EARTH: 600 KM (373 MILES)

SPACE TIMELINE

HUMANS DID NOT BEGIN to explore space until the twentieth century. In the beginning, small unmanned rockets and satellites were used. Since then spacecraft have developed into the most complex and expensive equipment ever built. Here are some milestones in space history.

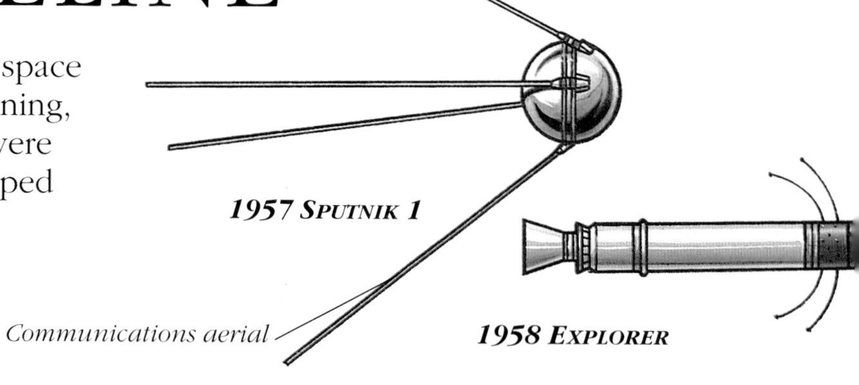

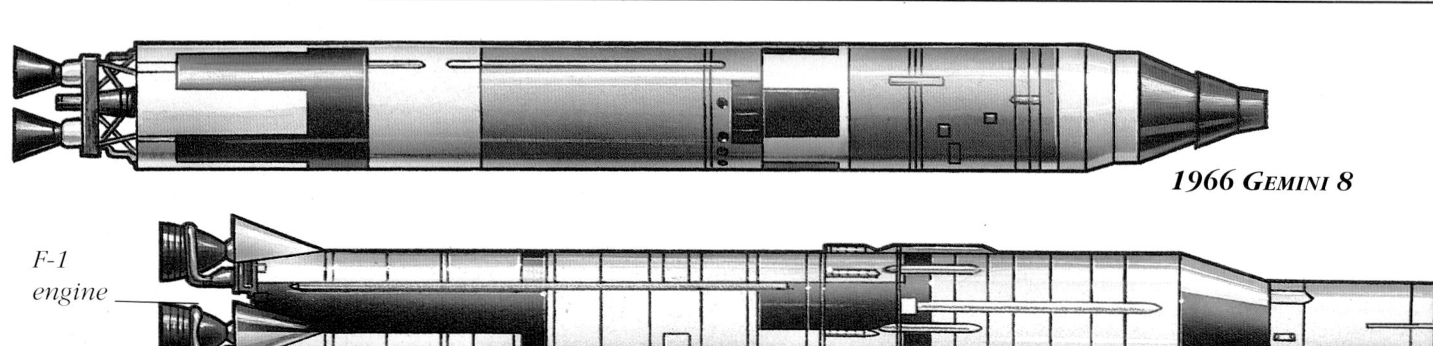

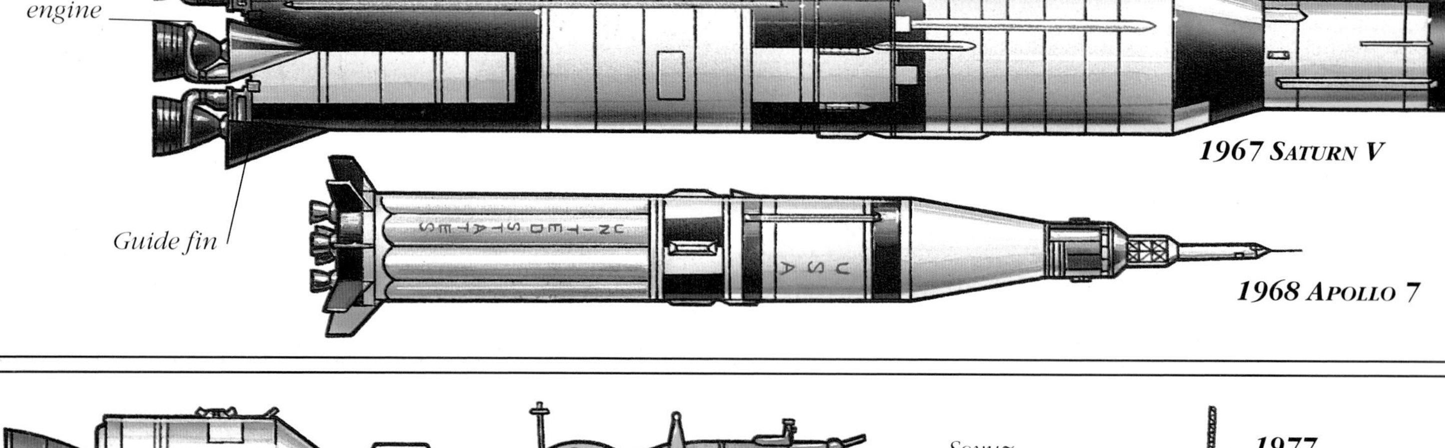

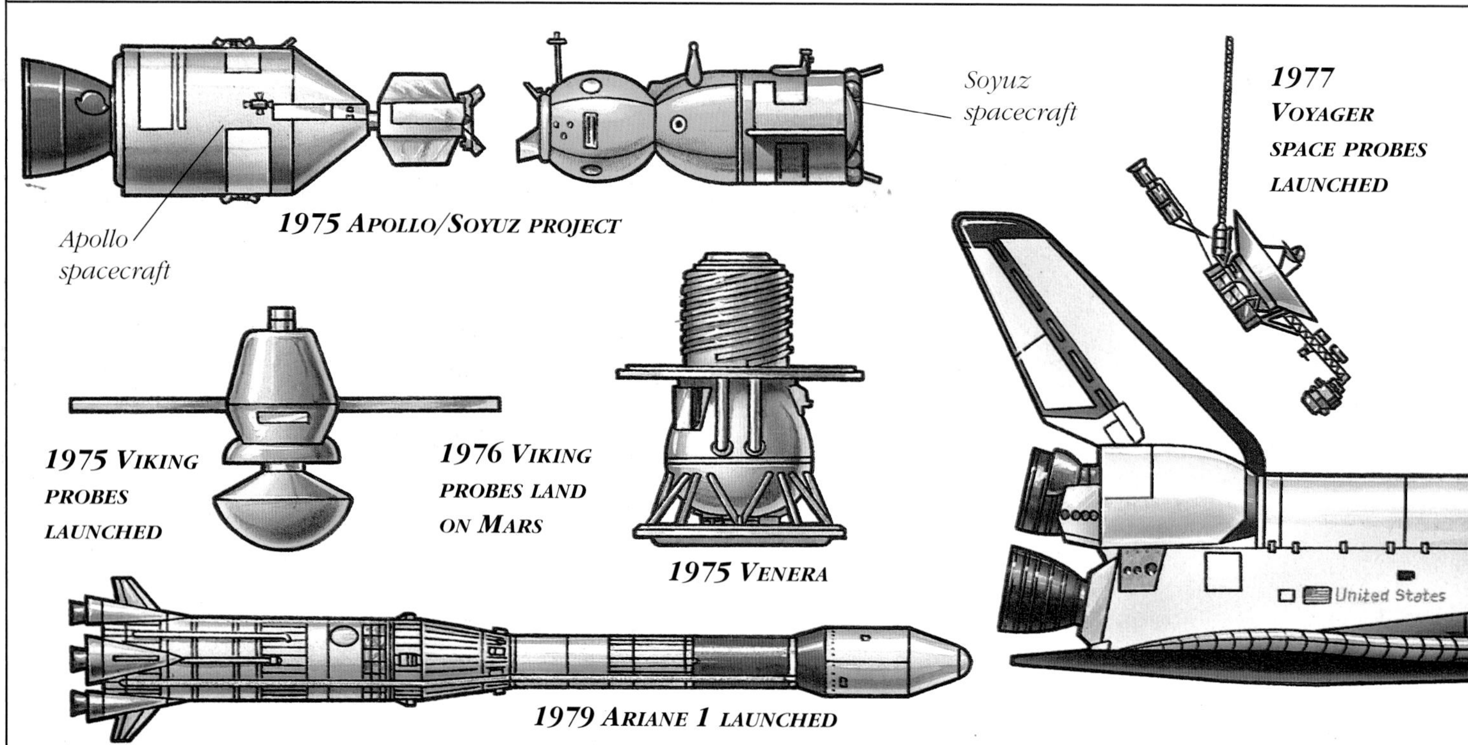

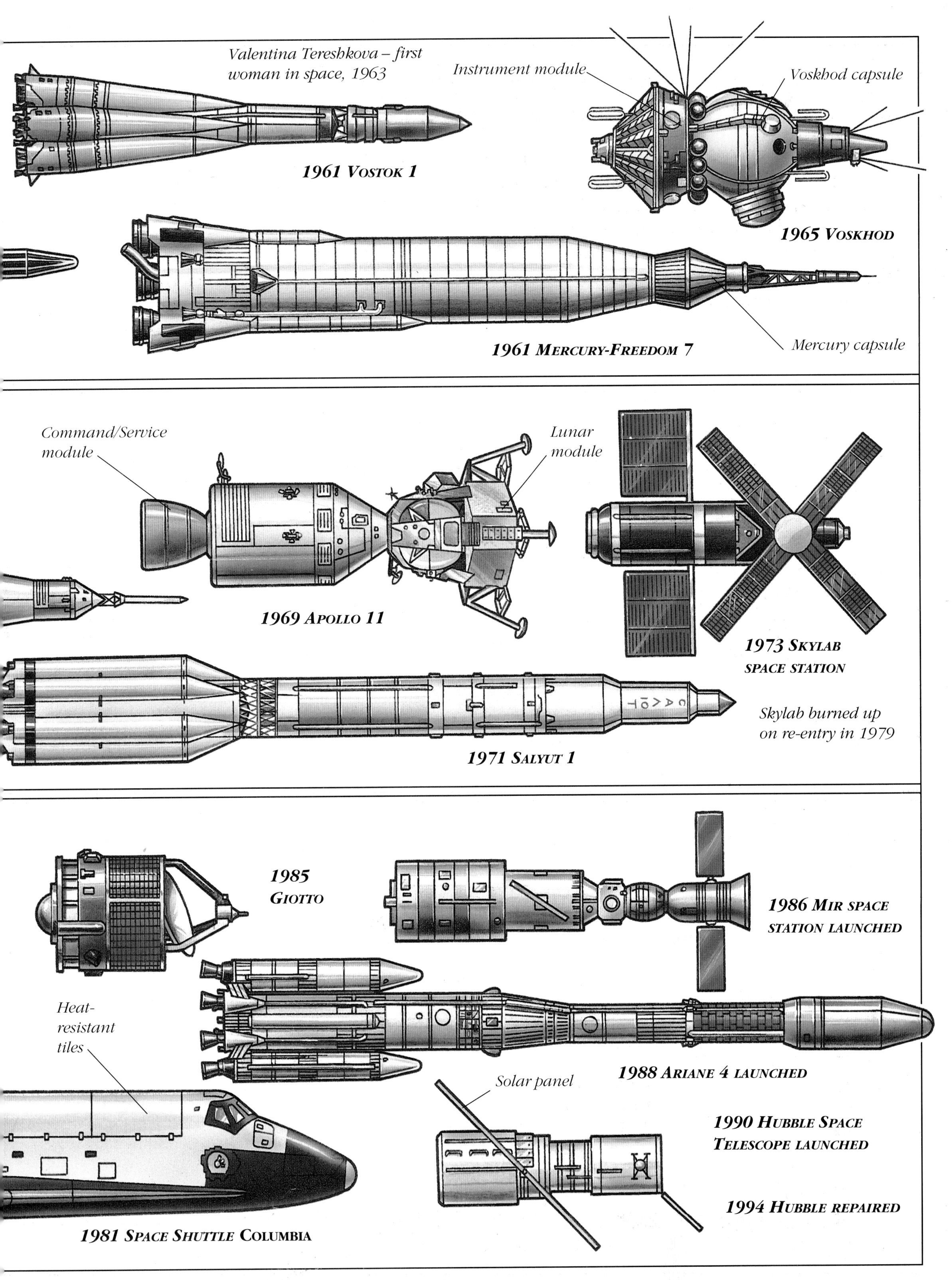
Valentina Tereshkova – first woman in space, 1963
Instrument module
Voskhod capsule
1961 VOSTOK 1
1965 VOSKHOD
1961 MERCURY-FREEDOM 7
Mercury capsule
Command/Service module
Lunar module
1969 APOLLO 11
1973 SKYLAB SPACE STATION
Skylab burned up on re-entry in 1979
1971 SALYUT 1
1985 GIOTTO
1986 MIR SPACE STATION LAUNCHED
Heat-resistant tiles
1988 ARIANE 4 LAUNCHED
Solar panel
1990 HUBBLE SPACE TELESCOPE LAUNCHED
1994 HUBBLE REPAIRED
1981 SPACE SHUTTLE COLUMBIA

Glossary

Airlock
A chamber in a spacecraft with an inner door connecting with the pressurized cabin and an outer door leading to space. Crew members usually put on spacesuits here. If they didn't use an airlock, all the air in the craft would be sucked out into space when they went outside.

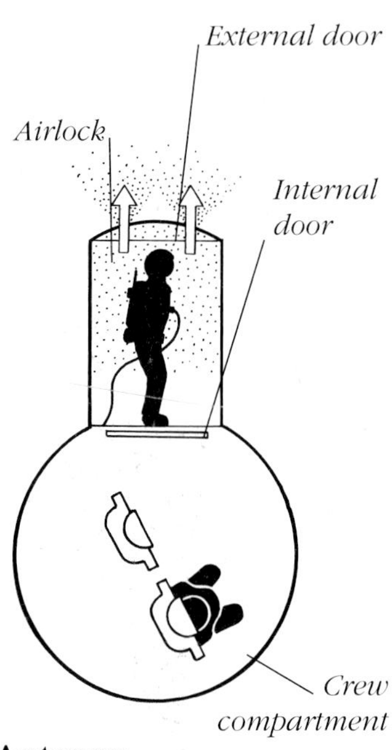

Antenna
A dish or rod aerial for receiving and sending radio signals to and from Earth.

Booster rockets
Rockets fitted on to a bigger launch rocket to help it gain extra speed as it travels up into space.

Command and Service Modules
CSM for short. Part of an Apollo spacecraft. It orbited round the Moon with one crew member on board while the other crew members landed on the Moon's surface in the Lunar Module.

Console
An instrument panel with controls and displays for a space crew to use.

Delta wing
A swept-back wing shaped like a giant V, used on the Space Shuttle orbiter.

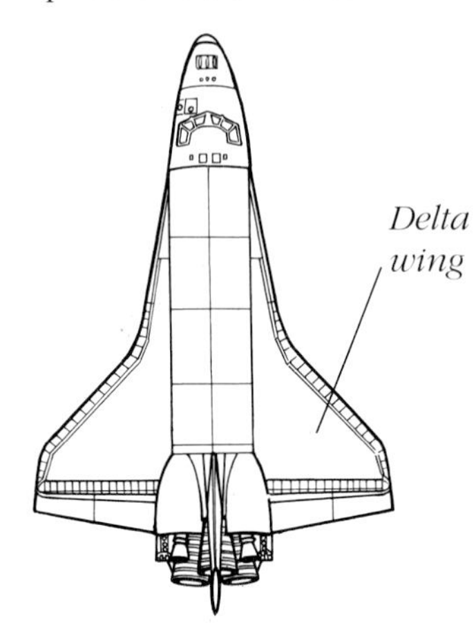

Depressurized
A place is depressurized when all the air is removed from it. For instance, airlocks are depressurized when astronauts are ready to go outside on a spacewalk.

Docking
One spacecraft joining up with another in space.

Docking adapter
The part of a spacecraft designed to lock on to another spacecraft when they dock together.

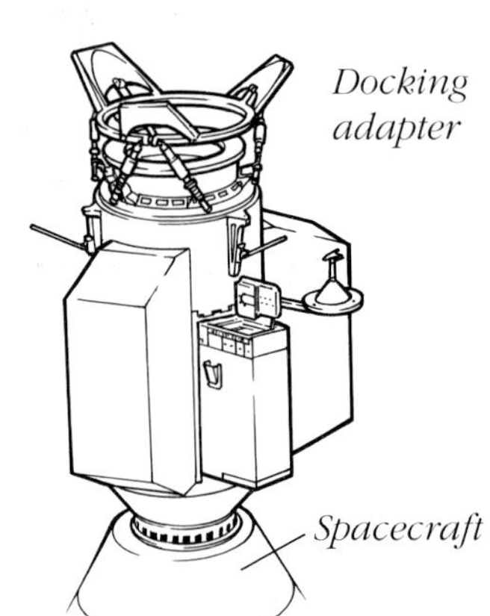

Docking hatch
A hatch which can be opened between two docked spacecraft, so that crew members can move through from one to the other.

Extra-vehicular mobility unit
EMU for short. Space jargon for a spacesuit.

Extra-vehicular activity
EVA for short. Space jargon for a space walk.

Fuel
The substance needed to make rocket engines work. Some fuel is liquid, some is solid and rubbery. It is burned together with a substance called an "oxidizer" to make gases that rush out of engine nozzles, pushing a launch rocket or spacecraft forwards.

Heat shield
A protective layer of heat-resistant material built round a spacecraft. This is particularly important if a manned spacecraft returns to Earth. As it plunges down, the outside surfaces get very hot and the crew need to be protected inside their cabin.

Life-support system
Equipment that provides crew members with the air, water, and warmth they need to survive in space.

Lunar Module
LM for short. The part of the Apollo spacecraft that landed on the Moon.

Lunar Roving Vehicle
LRV for short. A battery-powered vehicle used for driving over the surface of the Moon.

Mission Control
The main space centre on Earth where scientists monitor a spacecraft and keep in contact with the crew members on board.

Manned Manoeuvring Unit
MMU for short. A jet-powered backpack used by astronauts to fly around outside their spacecraft.

NASA
The National Aeronautics and Space Administration. The organization in charge of space exploration on behalf of the United States, founded in 1958 by President Eisenhower.

Nosecone
The top part of a launch rocket. Manned spacecraft or satellites sit inside the nosecone while they are being taken up into space.

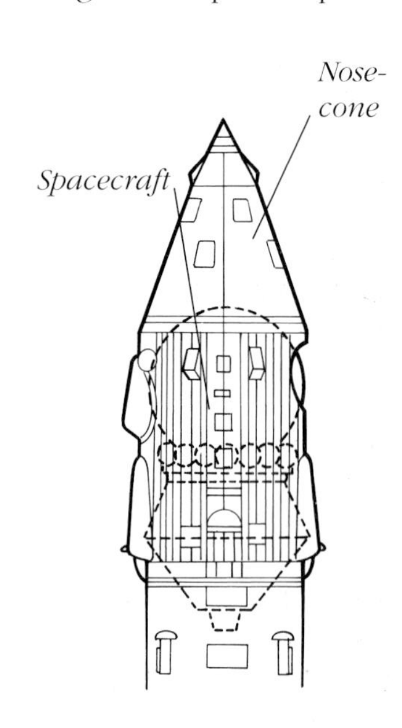

Orbit
A circular path followed by a small object circling round a larger object, such as a satellite circling round the Earth or an Apollo spacecraft circling round the Moon.

Orbit

Oxidizer
A substance (usually a gas) that is burnt together with fuel to drive a rocket engine.

Payload
A commercial cargo, such as a satellite, carried on board a spacecraft. Customers pay to send it into space.

Personal hygiene station
The bathroom/toilet on board a spacecraft.

Pressure suit
A simple form of spacesuit sometimes worn inside the cabin of a spacecraft. It protects crew members in case the cabin loses its air supply during a critical part of the mission, such as launching or landing.

Pressurized
A place is pressurized when it is filled with air. Spacecraft cabins and spacesuits are pressurized to imitate the pressure existing in the atmosphere around the Earth.

Reaction-control system
Controls, usually mini rocket nozzles, which are used to change a spacecraft's position.

Re-entry
The point when a spacecraft re-enters the Earth's atmosphere. At this stage air molecules start to rub against the craft as it falls, making its outer surface very hot.

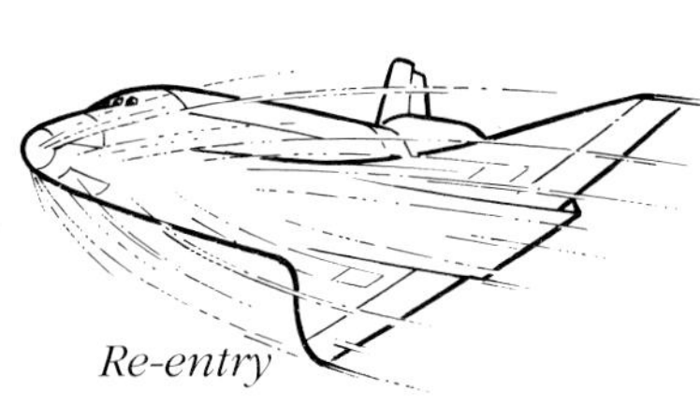

Re-entry

Remote Manipulator System
RMS for short, also called Canadarm. The robot arm attached to a Space Shuttle. It is used for jobs such as launching and repairing satellites. It was made in Canada.

Rocket
An engine that carries its own fuel and oxygen so that it can work in space as well as in the atmosphere. It is pushed upwards by gases streaming out of its exhaust nozzles. Launch vehicles are made up of several rocket stages linked together.

Remote sensing instruments
Scientific instrument fitted to spacecraft to do jobs such as collecting information about planets and stars.

Satellite
An object that circles (orbits) round a much larger body. Artificial space satellites are unmanned. They orbit the Earth doing different jobs, such as relaying telephone calls and reporting on the weather.

Satellite

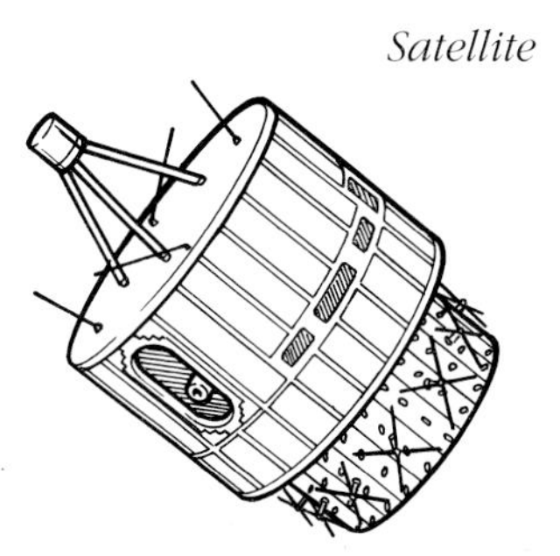

Sleep station
The crew sleeping quarters on board a manned spacecraft.

Solar array
A panel covered in a sheet of solar cells. These collect sunlight and convert it into electricity which can be used to run equipment on a spacecraft.

Space probe
An unmanned spacecraft sent to gather information about other planets and stars. Some space probes land, such as Viking. Some, such as Voyager, fly past planets, collecting information as they go.

Space station
A manned spacecraft designed to orbit the Earth for a long period. Crew members can live and work on board for periods of several months.

Spacelab
A laboratory workshop designed to fit in the payload bay of the Space Shuttle.

Splashdown
The moment when a manned spacecraft hits the water, if it splashes down in the ocean on its return to Earth.

MERCURY SPLASHDOWN

Parachute

Air-filled skirt cushions impact

Recovery helicopter

Splashdown

INDEX

Acknowledgements

Dorling Kindersley would like to thank the following people who helped in the preparation of this book:

Lynn Bresler for the index
Additional artworks by Brihton Illustration